T0199398

Proceedings of the 30th International Geological Congress
Volume 12

Palaeontology and Historical Geology

Proceedings of the 30th International Geological Congress

PROCEEDINGS OF THE
30TH INTERNATIONAL GEOLOGICAL CONGRESS

BEIJING, CHINA, 4 - 14 AUGUST 1996

VOLUME 12

PALAEONTOLOGY AND HISTORICAL GEOLOGY

EDITORS:
JIN YU-GAN
NANJING INSTITUTE OF GEOLOGY AND PALAEONTOLOGY, NANJING, CHINA
D. DINELEY
DEPARTMENT OF GEOLOGY, UNIVERSITY OF BRISTOL, BRISTOL, UK

CRC Press
Taylor & Francis Group
Boca Raton London New York

CRC Press is an imprint of the
Taylor & Francis Group, an **informa** business

First published 1997 by VSP BV Publishing

Published 2019 by CRC Press
Taylor & Francis Group
6000 Broken Sound Parkway NW, Suite 300
Boca Raton, FL 33487-2742

© 1997 by Taylor & Francis Group, LLC
CRC Press is an imprint of Taylor & Francis Group, an Informa business

First issued in paperback 2019

No claim to original U.S. Government works

ISBN 13: 978-0-367-44814-1 (pbk)
ISBN 13: 978-90-6764-257-6 (hbk)

Visit the Taylor & Francis Web site at
http://www.taylorandfrancis.com

and the CRC Press Web site at
http://www.crcpress.com

CONTENTS

Proc. 30ᵗʰ Int'l. Geol.. Congr., Vol. 12, pp.1
Jin and Dineley (Eds)
© VSP 1997

An Introduction to the Symposia on "Palaeontology and Historical Geology"

JIN YUGAN

Nanjing Institute of Geology & Palaeontology, Chinese Academy of Science, Nanjing, 210008 China

In this disciplinary section, six symposia were held. The presentations of the symposium on "Palaeobiogeography and reconstruction of palaeocontinents" (Chairmen, Jin Yugan and A. M. Ziegler) covers a wide range of topics from the Ordovician to the Miocene. But the Permian world is the main focus of the symposium. Twelve out of twenty-one presentations dealt with the newly developed continental reconstruction, climate and biogeographic models, and versions of the end-Permian extinction which provide a useful framework for understanding the enigma of Pangea evolution.

The symposium on "Global mass extinction and subsequent biotic recovery through geological history" (Chairmen: Douglas H. Erwin, A. Hallam and Hao Yichun) was organized in co-operation with IGCP project 335. Papers on the coal gap, coral gap and the possible preservation bias to geological record contributed new insights for thinking the delay of recovery after end-Permian extinction.

As in the previous congresses, a symposium on "Palaeocommunities through geological history" (Chairmen: A.J. Boucot, D. L. Bruton and Chen Yuanren) was scheduled for the 30th IGC. From fifty-three abstracts submitted, twenty four were presented to the symposium. Presentations with new and diverse data indicate a strong and continuing interest in synecological analysis of fossils and diverse geological application.

During the symposium on "Taphonomy, trace fossils, and extraordinarily preserved fossil groups (Chairmen, P. Crimes, C. G. Maples and Wu Xiantao), a majority of eighteen presentations covered various aspects of trace fossils and the lagerstattens.

The symposium on "Evolution of marine vertebrates", chaired by Zhang Meeman , was a successful meeting with the presentation of 4 papers. The relatively small scale of the symposium promoted fruitful comments and observation from the participants.

The Symposium on "Evolution and Environmental Significance of Calcareous-algae, Stromatolites and Mud-mounds" (Chairmen: R. Riding and Zhu Shixing) was held with support from the IGCP Project 380. Papers presented reflect a growing interest in the investigation of automicrites, and the possible microbiogenetic deposits.

We acknowledge the following scientists that acted as referees of the papers presented for publication in this volume: Professors A. J. Bocout, P. Crimes, B. Glenister, R. Riding and J. Utting.

Proc. 30ᵗʰ Int'l. Geol.. Congr., Vol. 12, pp. 2-17
Jin and Dineley (Eds)
© VSP 1997

A Hierarchical Framework of Permian Global Marine Biogeography

T. A. GRUNT[1], G. R. SHI[2]

[1]*Palaeontological Institute, Russian Academy of Sciences, Moscow, Russia*
[2]*School of Aquatic Science and Natural Resources Management, Deakin University, Rusden Campus, 662 Blackburn Road, Clayton, Victoria 3168, Australia*

Abstract

A hierarchical framework comprising 3 realms, 8 regions and 31 biotic provinces is proposed for the Permian global marine biogeography. The three realms, namely Gondwanan (Anti-Boreal or Notal), Palaeo-Equatorial and Boreal, are considered to largely correspond to, the southern polar to subpolar, palaeo-tropical and northern polar to temperate climatic zones, respectively. Regions are recognized within realms mainly on the basis of significant palaeogeographical barriers, both oceanic and continental. The Palaeo-Equatorial Realm is thought to be composed of the Cathaysian (located in the Eastern Tethys), Mediterranean (located in the Western Tethys), North American Region, and the Cimmerian Regions (located in the Southern Tethys). The Cimmerian Region displayed a highly dynamic nature in terms of provincial affinities with the Gondwanan and Palaeo-Equatorial Realms during the Permian, in that it belonged to the Gondwanan Realm in the Asselian-Tastubian time, then went through a mid-Permian transitional stage and finally entered the Palaeo-Equatorial Realm in the Late Permian (Dzhulfian to Changhsingian). The Boreal Realm is divided into two regions, Euro-Canadian Region and Taimyr-Kolyma Region, roughly corresponding to the northern palaeo-subpolar or palaeo-temperate and high palaeo-polar zones, respectively. Similarly, two regions are recognized for the Gondwanan Realm: Australian and Afro-South American. Discussions are focused on realms and regions, with comments also made to some provinces wherever appropriate.

Keywords: Permian, marine biogeography, hierarchy

INTRODUCTION

In many respects the Permian Period is uniquely placed in the geological history. It is a period of frequent and, at times, abrupt climatic changes and continental re-organization. The widely accepted concept of a Permian supercontinent, Pangea, has led some people to believe, perhaps erroneously to some extent, that the Permian world must have been very different from what we see today. While this perception may be true in terms of the geometry and number of continents, the Permian and the modern world share many other physical parameters. For instance, Permian landmasses, as depicted by most Permian palaeogeographical reconstructions, stretched from the South Pole to nearly the North Pole, a longitudinal configuration very similar to that of today. Several recent numerical climatic models based on a perceived, idealized Permian supercontinent have also predicted a pronounced differentiation of latitude-parallel climatic zonation that is comparable to that of the modern world [7, 17].

Thus, it is not surprising to see that palaeobiogeographers have also recognized a crude comparison of the Permian marine biosphere to the provincialism of modern seas in that three distinct, broadly latitude-parallel realms existed in the Permian, as they do at the present. These three Permian first-order biogeographical units (or biochores), hitherto widely known as the Boreal, the Tethyan or Equatorial and the Gondwanan, have been long recognized, but the spatial and temporal positions of their mutual boundaries and the number and nature of their constituent biochores remain to be defined. Over the past twenty years, a suite of more than 50 provincial names have been proposed to identify, in most cases with respect to only limited or local regions or areas, particular marine faunas each characterized by a varied number of taxa - species, genera or families. The proliferation of these names have caused nomenclatural confusion and sometimes misunderstanding. It has become increasingly apparent that some provincial names are clearly synonymies of others and that inconsistency exists in the ranking of Permian biogeographical units.

This paper is thus aimed to review existing Permian biogeographical schemes in an attempt to consolidate and define an unified and, in our opinion, a consistent Permian global marine biogeographical framework. In proposing this new unified scheme, we draw the palaeontological data primarily from Brachiopoda as this is a faunal group with, undoubtedly, the widest palaeogeographical distribution and greatest ecological tolerance and as the authors have first-hand knowledge on the systematics, biostratigraphy and palaeobiogeography of this group. However, wherever appropriate, data from other fossil groups will also be incorporated in the discussions. In presenting the framework, we place our focus on the distribution and characteristics of realms and regions, with less attention paid to the provinces as the latter would require a much lengthier paper. A point that has been avoided for discussion in this paper is on the usage of the Permian time scale. We have adopted the 'traditional' Tethyan time scale with a view in mind that an adoption of an alternative time scale would not alter significantly the overall provincial structure of the Permian marine biotas recognized in this paper. However, we acknowledge that significant progress has been made to consolidate and unify the Permian time scale in recent years under the International Subcomission on the Permian Stratigraphy and that the problem of reaching such a scale is still being widely debated (see papers presented in [35]).

THE CONCEPT OF HIERARCHY IN BIOGEOGRAPHY

As the hierarchical approach to the classification of provincial patterns has been as widely used in biogeography as it has in taxonomy, it is necessary here to give a brief account on the logic and significance of this approach in our analysis and classification of the Permian global marine provincialism.

Hierarchies exist almost in all natural systems in which units or entities are related to each other by ranking and ranks. Two general types of hierarchies have been recognized in biology and palaeontology: constitutive hierarchy and aggregative hierarchy [36]. The latter is of particular interest herein because it accommodates the hierarchy of biogeographical units, namely realms, regions, and provinces. In this aggregative ecological hierarchy, the province is the fundamental unit, and provinces of equal rank are nested to form a region, and regions of equal rank are amalgamated to form a realm.

Recognition of hierarchies in biogeography is not only a convenient and useful approach to summarize and simplify the complexity of the distribution of biotas in space; perhaps more importantly, it reveals the existence of natural hierarchies in the biogeographical (ecological) systems. Using floristic data, McLaughlin [20] has statistically tested for the existence of natural hierarchies in the distribution of floristic areas. A similar result was also obtained by Shi and Archbold [28] in their study of early Early Permian circum-Pacific brachiopod biogeography.

Like many hierarchies in biology, the formation of biogeographical hierarchies is believed to be related to parallel structures within the environments. Spatial heterogeneity of the environmental system is functional at all but varied spatial scales and appears to be hierarchically structured [1, 22]. This may be exemplified by an ecosystem within a large marine basin, such as the Pacific. Differential reception of solar energy and associated differential atmospheric circulation patterns divide the basin into several clearly definable, large-scaled, roughly latitude-parallel climatic zones (belts). Within each of these zones, zonal winds cause circulation patterns in the basin that divides it into distinct hydrogeographical regions. Within each such region, which on a global scale may be regarded as relatively homogeneous, smaller-scale environmental heterogeneity, such as upwelling, fronts, eddies, seasonality and continental shelves, may be generated by tidal and regional wind forcing and interactions of tides and other forms of water body movements with localized bottom topography or continental margin. It is this hierarchically structured environmental system that we believe has induced similar hierarchical patterns of biotic distributions within many of the large modern ocean basins.

CLIMATIC ZONES, GEOGRAPHICAL BARRIERS AND DIFFERENTIATION OF BIOCHORES

As discussed in the preceding section, the hierarchy of biogeographical units is not only a practically convenient and useful approach to analyze and classify the otherwise complex biospheric systems, it is perhaps more a natural revelation of the true structures of the ecosystems influenced upon and ultimately determined by the hierarchy of environmental parameters. According to this principle, we propose to classify the spatial distributional patterns of Permian marine biotas in a hierarchical framework of, in ascending order, provinces, regions, and realms. Subordinate unites are possible but are not adopted here because they are generally more difficult to define and justify, especially in relation to the adjacent units in the biogeographical hierarchy.

Thus, a realm is regarded as the first-order biogeographical unit distinguished on a global scale by the presence (or absence) of endemic orders, superfamilies, and/or families. Although in theory recognition of realms is not based on large-scale climatic zonation, studies of modern biogeography (e.g. [8, 23, 25]) and palaeobiogeography (papers in Hallam [13] and McKerrow and Scotese [19]) have demonstrated a distinct correlation between major realms, modern or ancient, with (palaeo)climatic zonation patterns, suggesting the primary control of latitude-related climates and hence temperature gradients on a large-scale global distributional patterns of organisms. The distribution of some sedimentary build-ups, such as reefs and bioherms, could be also used for the identification of ancient climatic zones (belts).

Within each realm, areas of endemic or characteristic families, subfamilies and genera may be recognized as a second-order biochore in the hierarchy of biogeography. A biogeographical region may be characterized as a vast aquarium confined within the limits of one realm or a latitudinal climatic zone and separated from other aquaria by environmental barriers. Among these environmental barriers, which may be biotic or abiotic, geographical separation appears to play a primary role in the regionalisation of biotas within a realm. In some cases, geographical barriers can be even as significant as climatic zones in affecting marine provinciality, depending on the size and orientation of the barriers. For shallow marine benthos, large landmasses and deep ocean basins can be effective biogeographical determinants and commonly form regional as well as provincial boundaries.

Continued development of a fauna within a large region (or a single marine basin) over a prolonged geological period may also facilitate the formation of a biogeographical region characterized by the independent, historically inherited development of a large number of taxa. The North American biogeographical region, for instance, may be cited as a good example of this kind of biogeographical development. This biogeographical region, developed in a platform-type marine basin within the palaeo-tropical zone, continued to manifest as a distinct regional-rank biochore from the Ordovician to the early Late Permian.

Regions may be divided into lower-ranked biochores, widely used as provinces. We regard a biotic province, much like the species concept in taxonomy, as a fundamental biogeographical entity and define it as a spatially contiguous area located within a region and inhabited by a characteristic association of species and endemic genera in some cases. Ideally, provinces should be identified on the basis of species, preferably as species diversification centers. This approach has been widely adopted by neobiogeographers but its significance in palaeobiogeography has been variably treated. It is argued that the selection between genera or species as the taxonomic operational units (OTUs) of a palaeobiogeographical study is perhaps more a matter of the spatial scale in use. In many previous studies, genera have been proved to be effective OTUs in inter-continental or global palaeobiogeographical studies, especially statistically based analyses (e.g. [6, 28]), just as species have also been proved effective in detecting regional and continental provincial patterns (e.g. [11]). In general, palaeobiogeographical studies based on species tend to identify more provinces than if genera are used. In order to avoid potentially overestimating or underestimating provinciality of a given geological time, a compromising strategy may be adopted, combining both genera and species, or a selected set of characteristic genera and species (e.g., [2]).

In summary, the biogeographical hierarchy does not exist in isolation; it can be closely correlated and may therefore be assumed to be linked to the hierarchies of taxonomy and the environments. These inter-relationships may be graphically presented as follows:

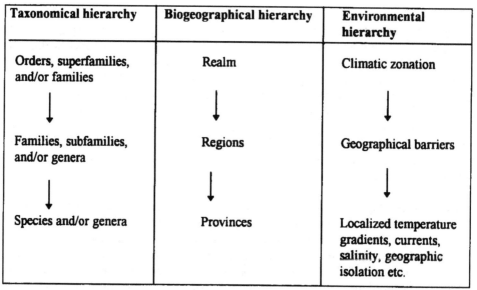

Taxonomical hierarchy	Biogeographical hierarchy	Environmental hierarchy
Orders, superfamilies, and/or families ↓	Realm ↓	Climatic zonation ↓
Families, subfamilies, and/or genera ↓	Regions ↓	Geographical barriers ↓
Species and/or genera	Provinces	Localized temperature gradients, currents, salinity, geographic isolation etc.

Figure 1. Correlation between the biogeographical hierarchy with taxonomical and environmental hierarchies (see text for more discussion).

PERMIAN GLOBAL MARINE BIOGEOGRAPHICAL FRAMEWORK

Adopting the hierarchical concept discussed above, we divide the global Permian marine basins into 3 realms, 8 regions and 31 biotic provinces (Fig. 2). In this scheme, we have retained most of the existing provincial names wherever appropriate. For the sake of page limits, references to the origin of these provincial names are omitted but are available from the authors.

This framework is a summarized and revised version of two existing schemes recently independently proposed by us [12, 27] and represents an alternative interpretation to that of Bambach [6]. The latter study, largely based on binary (presence/absence) similarity analysis, recognized 4 realms, 4 regions (all from one realm) and 17 provinces for the Early Permian, and the same 4 realms, 4 regions but 15 provinces for the Late Permian. The difference between our framework and that of Bambach at the realm level is superficial and could be regarded as an illusion caused by ranking. Two of Bambach's realms, namely Tethyan and American Realms, are herein downranked

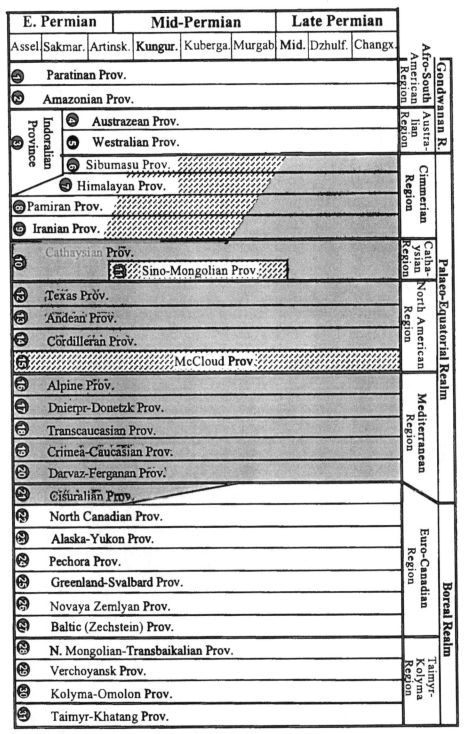

Fig 2. A hierachical classification of Permian global marine realms, regions and provinces. (The shaded area indicate the extent of the Palaeo-Equatorial Realm, and the stippled areas are transitional provinces of mix marine faunas).

Figure 3. Asselian–Tastubian (early Early Permian) global marine provinces plotted on a Sakmarian global reconstruction map of Ziegler et al. [39] with modification on the palaeopositions of Cathaysian continents (numbers correspond to the identification numbers of provinces in Fig. 2)

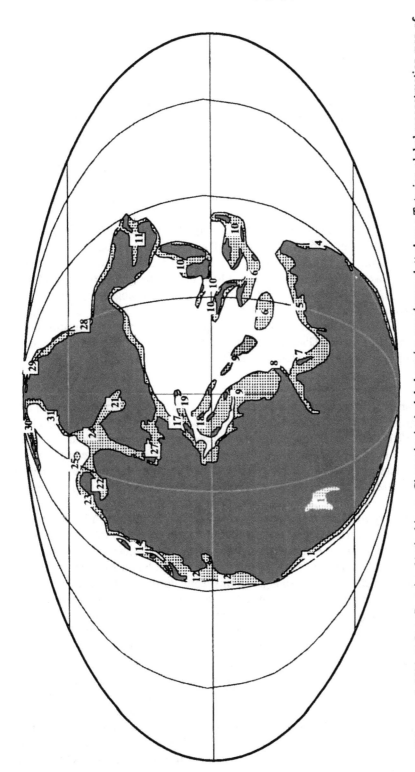

Figure 4. Mid- to Late Permian (Artinskian to Changxingian) global marine provinces plotted on a Tatarian global reconstruction map of Ziegler et al. [39] with minor modification on the palaeopositions of Cathaysian and Lhasa microcontinents and an addition of a seaway into northeast China (numbers correspond to the identification numbers of provinces in Fig. 2).

as regions within the Palaeo-Equatorial Realm because they are believed to have been located in the same or similar palaeo-latitudinal (and hence palaeoclimatic) zones, on or near the Palaeo-Equator, therefore justifying to classify them together within one broad palaeo-tropical realm as defined previously.

An examination of the temporal distributions of the Permian biotic provinces (Figs. 2-4) reveals two features of particular interest. Firstly, some provinces only existed for a relatively short period of time during the Permian. For instance, the distinct Indoralian Province was most prominent during the height of the Asselian-Tastubian Gondwanan glaciation. The breakup of this biochore into a number of provinces in the post-Tastubian times coincided with the deglaciation of Gondwana, reflecting the close control of climatic conditions on provincialism. Another interesting feature of the hierarchical framework is the dynamic nature of the boundaries between the biotic provinces, regions and realms. It is clear from Fig. 2 that some provinces and regions changed their identities during the Permian. For instance, the Cisuralian Province of the Mediterranean Region belonged to the Palaeo-Equatorial Realm in the Early Permian, then went through a transitional phase during the mid-Permian, and finally moved to the Boreal Realm in the Late Permian. A similar biogeographical signature can be also identified for the Cimmerian Region, which aligned itself with the Gondwanan Realm during the early Early Permian, then through a transitional stage characterized by a mixed Gondwanan and Cathaysian fauna, and finally joined the Palaeo-Equatorial Realm in the late Late Permian.

The Palaeo-Equatorial Realm

Various names have been proposed for more or less the same geographical extent of the Palaeo-Equatorial realm; among them the most widely used is Tethys or Tethyan Realm [e.g. [10, 30]). The term Tethys is not followed here because it was originally proposed and has since been widely used as a palaeogeographical entity (a marine basin) wedged between the Eurasia in the north and Gondwana in the South. As will be seen later, this large marine basin actually spanned across all the three realms during the Permian.

On a modern geographical map, areas covered by this realm occurs in low to moderate latitudes, extending from central and southern Europe, eastwards through the Middle East, central and southern Asia to East and Southeast Asia; westwards it occupies much of southwest Unites States, Central America and northern South America. In addition, there are also fragments which belonged to this realm during the Permian but had been rifted away and accreted to higher latitudinal regions as allochthonous terranes [14]. There are three key lines of evidence to support the interpretation of the realm as of palaeo-equatorial (hence palaeo-tropical) origin. Firstly, palaeomagnetic data obtained from Permian rocks within various parts of the realm have indicated a low, ranging generally between 0 to 20°S or N, palaeo-latitudinal setting (see palaeomagnetic data compilations in [13, 37]. In association with and support of the palaeomagnetic data, sedimentological and palaeontological evidence from the Permian rocks of the supposed palaeo-equatorial areas also suggest low, palaeo-tropical conditions. These evidence include high biotic diversity, wide-spread distribution of reefal buildups, and red beds and other evaporitic deposits (halites, gypsium), all suggesting warm-water or arid conditions. A third, more direct, line of evidence has been derived from isotopic studies of organic shells from many parts of the 'palaeo-equatorial belt', which have also pointed to the prevalence of a warm, tropical condition for the realm throughout the Permian [38].

Biogeographically, the realm is typified by the presence of several endemic higher taxa thermally adapted to warm-water conditions; they include verbeekinid fusulinids, waagenophyllid corals, and richthofeniid and oldhaminoid brachiopods. With the exception of scattered occurrences of fusulinids in the Arctic region and other isolated occurrences in some displaced terranes, fusulinaceans are almost confined to the Palaeo-Equatorial Realm. The same has also been observed to be true for the compound rugose corals. Reefal buildups are also widespread within the realm but absent or found very rare from the Gondwanan or Boreal biotas. Despite these common features, the Palaeo-Equatorial Realm is internally heterogeneous, which makes it possible to delineate further biogeographical subdivisions. Accordingly, three regions are recognized, herein named the Cathaysian Region, North American Region and the Mediterranean Region. An additional region, the Cimmerian Region, also became a part of the Palaeo-Equatorial Realm towards the Late Permian following a transitional mid-Permian stage (Fig. 2).

The differentiation of the Cathaysian, North American and Mediterranean Regions within the Palaeo-Equatorial Realm is best explained by geographical separation and sufficiently prolonged independent development [10]. As has been discussed, these three regions were located broadly on or near the palaeo-equatorial zone during the Permian and hence confined within the same palaeo-tropical climatic belt, but they were separated from each other by two, then the largest, ocean basins (Tethys and Panthalassa) and a continental mass (Pangea). The Panthalassa separated the Cathaysian Region from the North American Region, while the Tethys sea isolated the Cathaysian Region from the Mediterranean Region, and the supercontinent Pangea demarcated the Mediterranean Region from the North American Region.

The difference between North American and Euro-Asian warm-water faunas has long been noticed. Bambach [6] regarded this difference very highly and accordingly classified them into two different realms. We differ from this view. We believe that realms as first-order biochores are primarily controlled by large-scale climatic zonations and hence faunas within the same realm are influenced by the same or similar climatic conditions. Regions, on the other hand, are believed to be more effectively controlled by geographical barriers and associated biological/ecological isolation. In view of these arguments, we consider the difference between the North American and Euro-Asian faunas to be a manifestation of regional differentiation. Because North American and Tethyan regions were both located in comparable palaeoclimatic settings, they are similarly characterized by the abundance of reefal structures, high diversity, and co-occurrences of waagenophyllid corals, verbeekinid fusulinids, and lytoniid and richthofeniid brachiopods. On the other hand, the immense Panthalassa Ocean must have rendered the migration of most benthic species and numerous genera between the two regions impossible during the Permian. For example, more than 30% of the Permian brachiopod genera known from the Glass Mountains in the southwest United

States have not been found in the Euro-Asian warm-water aquatoria, presumably reflecting the effect of the Panthalassan oceanic separation.

The Permian Tethys is believed to be a large marine basin. The depth, width, and thus the oceanic or continental crustal nature of this basin is still a matter of debate. Biogeographically, three distinct faunas developed within the Tethys, corresponding respectively to the Cathaysian Region, the Mediterranean Region, and the Cimmerian Region. The Cathaysian Region is located in the eastern part of the Permian Tethys, consisting of a number of islands or micro-continents including Sino-Korean Platform, South China, Indo-China, the Tarim Basin, and the North Qiangtang block. Permian palaeontological data from this region supports a biogeographically cohesive fauna from all the micro-continents [9]. This fauna is characterized by a high diversity and an estimated endemism upto 30% at species level (estimation based on Brachiopoda data).

The palaeogeographical extent of the Cathaysian Region varied significantly during the Permian, in correspondence to the rapid changes of Permian climates, geography and tectonics. Overall, three stages of evolution have been recognized [10, 30]. In the early Early Permian (Asselian-Tastubian), the Cathaysian Region was connected with the Ferganian and Darvas-Transalai Provinces in the southwest of the Tethyan basin and with the Cisuralian Province located in the northwest of the Tethys. (Fig. 3), both belonging to the Medditerranean Region. These connections allowed a high percentage of species to be shared between the Cathaysian and Mediterranean Regions. However, endemism in the Cathaysian Region remained high during this time. For instance, of the 120 brachiopod species reported from western Guizhou of the South China block, 32 or 37% were endemic to the Cathaysian [18].

At the same time, the boundary between the Cathaysian Region and the Cimmerian Region of the Gondwanan Realm appears to be "sharp", with little faunal communication between the two regions. This sharp separation in the southern part of the Tethys corresponded to the climax of the Late Palaeozoic Gondwanan glaciation, a factor therefore assumed to be responsible for this biogeographical demarcation.

The second stage in the evolution of the Cathaysian Region roughly corresponded to the mid-Permian (Late Sakmarian to probably Murgabian-Midian). By this stage, the Uralian seaway connection had been closed, but another marine connection (or connections) was opened in the northeast of the Tethys, connecting the Taimyr-Kolyma Region of the Boreal Realm through Transbaikalia and Mongolia to the Cathaysian Region. This seaway, accompanied by a mid-Permian (Kungruian-Ufimian) trans-Arctic transgression, brought many typical Boreal elements into north and southeastern Mongolia, northeast China, Russian Far East and parts of Japan to intermingle with existing Cathaysian taxa, resulting in the formation of a highly distinctive mixed marine fauna of the Sino-Mongolian Province [30, 32]. Parallel to the development of the Sino-Mongolian mixed (or transitional) province, the Cimmerian Region was also experiencing a transformation in provinciality, changing from a typical Gondwanan fauna to a more warm-water type. This stage of provincial transformation was also characterized by an admixture in the same region of both Gondwanan and Cathaysian elements [30].

The third stage of the provincial development of the Cathaysian Region was marked by the expansion in area extent of the region through incorporating the two mid-Permian transitional provinces. Inclusion of the Sino-Mongolian Province (which also included the Russian Far East) into the Cathaysian Region came about because, by the beginning of the Murgabian, the Mongolian-Transbaikalian seaway connection (or connections) to the Taimyr-Kolyma Region had been closed and the global climate was ameliorating, subsequently extinguishing the presence of Boreal elements in the northern part of the Cathaysian Region. The progressive incorporation of the Cimmerian Region into the Palaeo-Equatorial Realm through mid- to Late Permian is of particular interest and could be explained by several possible mechanisms [30]. This process is most likely to have been caused by the interaction of both a post-glaciation global gradual warming and rapid northward drift of the Cimmerian micro-continents.

The Mediterranean Region was located along the western coast of the Tethyan basin and was also geographically proximal to the North American Region. In the Asselian, the Mediterranean Region showed an unusual faunal similarity with that of the southwestern United States [10, 12]. Yet, it is believed that throughout the Permian, the Mediterranean Region was completely isolated from the North American Region. As already mentioned, the faunal communication between the Mediterranean the Cathaysian Regions was evident during the early Early Permian, but became more limited in the middle and Late Permian times due to intensified orogenic movements and related regressions. The broad Tethyan basin may also have played a significant role in regionalizing the Cathaysian and Mediterranean faunas.

As already referred to, the development of the Cimmerian Region throughout the Permian was highly dynamic and coupled with the changing nature of the interface between the Tethys and Gondwana. Recent stage-by-stage statistical analysis of the Permian brachiopod faunas from the Shan-Thai and Himalayan Provinces of the Cimmerian Region in relation to those of the Gondwanan Realm and the Euro-Asian Palaeo-Equatorial basins has clearly demonstrated a rapid and progressive changing profile in the provinciality of the Cimmerian Region (see a summary paper in [5]).

The basins of the North American Region, throughout the Permian, were characterized by a very high degree of biogeographical isolation; the biota was highly endemic, with most faunal groups (brachiopods and fusulinids among them) being characterized by specific evolutionary rates. Within the North American Region, a distinct transitional province is to be found in the McCloud belt (Figs. 2, 4). This biochore is characterized by a fauna of mixed origin, incorporating elements suggesting affinities with the Cathaysian Region, the Grandian and Cordilleran Provinces of the North American Region, as well as with the Alaska-Yukon Province of the Euro-Canadian Region within the Boreal Realm [26]. Tectonically, the McCloud Province is classified with the Eastern Klamath terrane, origin of which remains controversial (cf. [21]). Shi [26], based on the remarkable mixed nature of the Permian marine faunas found from this terrane, suggested that it could be located in the northern palaeo-temperate zone between palaeolatitudes 30° and 50°, with geographical proximity to both East Asia, northeast Asia and western North America.

The Boreal Realm

This realm, as interpreted by Ganelin and Kotlyar (in [16]), comprises the Permian basins of the Arctic, northern Asia, western and central Europe, as well as some low latitudinal regions as in the Zechstein basin of western Europe and the Sino-Mongolian region and the Russian Far East in East Asia in the earliest Permian Within this realm, two regions appear to be distinguishable from both palaeontological and sedimentological point of views. The Taimyr-Kolyma Region is characterized by a highly endemic fauna coupled with limited glacial or glacigenic sediments [34], implying a possible high polar geographical setting. Except for the isolated occurrences of fusulinids found in displaced Cathaysian terranes in northeast Siberia, the Taimyr-Kolyma Region lacks any records of either fusulinids or compound rugose corals. The diversity of the Taimyr-Kolyma fauna is low and dominated by Brachiopoda, Bivalvia and Bryozoa. This higher taxonomic composition is very similar to that of the Australian Region of the Gondwanan Realm. Another distinct comparable feature with the Australian Region is that both regions share a number of genera which are absent or only found very rarely elsewhere. The existence of these taxa points to a long recognized but still a puzzling biogeographical phenomenon of the Permian Period [31].

In an analogy to the geographical subdivisions of the modern Arctic region, the Permian Euro-Canadian biogeographical region is best regarded as an equivalent to the subarctic zone or temperate zone. This region now comprises the Alasaka-Yukon region, the Canadian Arctic basins, central and western Europe, Greenland, Svalbard, the Russian Platform and the Urals. Palaeogeographically, the Permian basins of Central Europe and the Urals were probably located in subtropical zones and connected to the Tethys at the beginning of the Permian, resulting in a highly "Tethyanised" early Early Permian (Cisuralian) fauna. However, by mid-Permian (Artinskian), the seaway connections ceased to exist due to intensified orogenic movements. After the orogeny and through the mid-Permian, the Cisuralian Province shifted from a subtropical zone to a relatively high latitudinal setting (probably cool temperate zone), giving rise to the transformation from a predominantly warm-water Mediterranean fauna in the early Early Permian to a cool-water, Euro-Canadian type fauna (Fig. 2). The tectonic shift of the Cisuralian Province since the Artinskian is evident from the comparison of the four successive Permian palaeogeographical reconstructions provided by Ziegler et al. [39].

The Gondwanan Realm

This realm corresponds to the conventional configuration of Gondwana. Reviews of Permian marine palaeobiogeogreaphy of the realm has been carried out by Archbold [2, 3] and Runnegar [24], with a number of provinces identified. In general, two biogeographical regions may be recognized, herein named the Australian Region and Afro-South American Region. The former includes faunas from the Australian continent, New Zealand and, in the early Early Permian (Asselian-Tastubian), the Cimmerian Region as well. A similarity analysis of the Asselian-Tastubian brachiopods from this broad region has revealed the prevalence of a single biogeographical fauna, called the Indoralian Province, dominating the whole region. The timing of the Indoralian Province corresponded to the climax of the Gondwanan glaciation. Starting from the Late Sakmarian (Sterlitamakian), the Indoralian Province began to differentiate, probably in response to the climatic amelioration and invasion of warm-water elements [4]. Subsequently, a number of provinces emerged (see Fig. 2). Throughout the mid to Late Permian, both the Westralian and Austrazean Provinces maintained high endemism. In particular, the Austrazean Province was characterized by an Ingelarellinae fauna with some 7 genera and more than 30 species, most of which are endemic to this province. This high degree of endemicity reflects its persistently high palaeo-latitudinal position (higher than 60°S) and palaeogeographical isolation from other provinces and regions. In contrast, the Westralian Province, probably due to its relatively lower palaeo-latitudinal setting and palaeogeographical proximity to the Cathaysian and Cimmerian Regions, developed a less endemic fauna with a variable amount of Cathaysian and Cimmerian influence [4]. By the Dzhulfian time, typical palaeo-tropical warm-water brachiopod genus *Leptodus* had reached here [33], probably signaling a southward expansion of the palaeo-tropical belt or the northward drift of the entire Gondwana to lower palaeo-latitudes.

The differentiation of the Afro-South American Region from its eastern counterpart is a natural manifestation of geographical barriers. In view of the palaeogeographical reconstruction provided by Ziegler et al. [39] and others, there appear to have been no migration routes possible between the Australian and Afro-South American Regions. However, the two regions share a number of Gondwanan-endemic genera and some morphologically closely related species (see [2] for discussion). Without a seaway connection between the two regions (cf. [28]), the faunal links would be hard to explain. In comparison with the Australian-New Zealand Permian faunas, those of the Afro-South American Region remain under investigated, hampering a detailed reconstruction of the regional provincialism; nevertheless, two provinces are provisionally recognized here, Paratinan and Amozonian, largely corresponding to two separate deposit settings (Figs 3, 4).

CONCLUSIONS

The investigation into the Permian global marine provincial patterns is necessarily a complex and ongoing task, requiring regular updates on existing schemes based on new data available. Along these lines, the framework presented in this paper is by no means final, nor it represents a universal view on the problem. The significance of this scheme, however, is that it represents a major step towards a unified framework and an attempt to consolidate existing nomenclature and provincial names. The concept of hierarchy is used herein but its underlying significance in understanding the Permian global marine provincialism in relation to palaeoclimates, palaeogeography and tectonics needs further investigation. A reasonable amount of details have been given to the discussions on the differentiation and characterization of realms and regions; however, detailed elucidation on the provinces have been deliberately left out due to page limits and will be presented elsewhere.

Acknowledgments

This paper is a result of TAG's recent trip to Australia funded by an Australian Research Council grant to GRS. GRS is responsible for opinions expressed in the first three sections of the paper, while TAG and GRS are jointly responsible for the section on the global Permian marine biogeographical framework. We thank Professor N.W. Archbold for discussion and encouragement.

REFERENCES

1. T.F.H. Allen and T.B. Starr. *Hierarchy-Perspectives for Ecological Complexity*. University of Chicago Press, Illinois (1982).
2. N.W. Archbold. Permian marine invertebrate provinces of the Gondwanan Realm. *Alcheringa*, **7**, 59-73 (1983).
3. N.W.Archbold. Paleobiogeography of Australian Permian brachiopod faunas. In: *Brachiopods*, P. Copper and Jisuo Jin (eds). Balkema, Rotterdam, 19-23(1996).
4. N.W. Archbold and G.R. Shi. Permian brachiopod faunas of western Australia: Gondwanan-Asia relationships and Permian climate. *J. Southeast Asian Earth Sci.*, **11**, 207-215.
5. N.W.Archbold and G. R. Shi. Western Pacific Permian marine invertebrate palaeo-biogeography. *Aust. J. Earth Sci.*, **43**, 635-641 (1996).
6. R.K. Bambach. Late Palaeozoic provinciality in the marine realm. In: *Palaeozoic Palaeogeography and Biogeography*. W.S. McKerrow and C.R. Scotese (eds). *Geol. Soc. Lond. Mem.* **12**, 307-323 (1990).
7. E.J. Barron and P.J. Fawcett. The climate of Pangea: a review of climate model simulations of the Permian. In: *The Permian of Northern Pangea. 1. Paleogeography, Paleoclimates, Stratigraphy*, P.A. Scholle, T.M. Peryt and D.S. Ulmer-Scholle (eds), Springer-Verlag, Berlin. 37-52.
8. S. Ekman. Zoogeography of the Sea. Sidgwick and Jackson Limited, London (1953).
9. Z.J. Fang. A preliminary investigation into the Cathaysian faunal province. *Acta Palaeont. Sin.*, **24**, 344-348 (1985).
10. T.A. Grunt. The biogeography of brachiopods of the Permian Tethys. *Biull. Mosk, Obshch. Ispy. Prir. Otdel. Geol.*, 90-101 (1970).
11. T.A. Grunt. The biogeography of the brachiopod order Athyridida. *Palaeont. Zhur.*, 1989(2), 40-51 (1989).
12. T.A. Grunt. Biogeography of Permian marine basins. *Palaeont. Zhur.* 1995(4), 10-24 (1995).
13. A. Hallam. *Atlas of Palaeobiogeography*. Elsevier, Amsterdam (1973).
14. A. Hallam. Evidence of displaced terranes from Permian to Jurassic faunas around the Pacific margins. *J. geol. Soc. Lond.*, **43**, 209-216 (1986).
15. E. Irving. Fragmentation and assembly of the continents, mid-Carboniferous to Present. *Geophys. Surveys*, **5**, 299-333 (1983).
16. G.B. Kotlyar and D.L. Stepanov (eds). Main features of Stratigraphy of the Permian System in the U.S.S.R. *Vses. ord. Lenina Nauchno-Issled. Geol. Inst., Trudy*, n.s., **286**, 1-272.
17. J.E. Kutzbach. Idealized Pangean climates: sensitivity to orbital change. *Geol. Soc. Am. Spec. Pap.*, **288**, 41-55.
18. L. Li, D. Yang and R. Feng. The brachiopod and the boundary of Carboniferous-Permian in Longlin region, Guangxi. *Bull. Yichang Inst. Geol. Miner. Resources*, **11**, 239-276(1987).
19. W.S. McKerrow and C.R. Scotese (eds). *Palaeozoic Palaeogeography and Biogeography*. *Geol. Soc. Lond. Mem.*, **12** (1990).
20. S.P. McLaughlin. Are floristic areas hierarchically arranged? *J. Biogeog.* **19**, 21-32(1992).
21. M.M. Miller. Dispersed remnants of a northeast Pacific fringing arc: Upper Paleozoic terrane of Permian McCloud faunal affinity, western U.S.A. *Tectonics*, **6**(6), 807-830 (1987).

22. R.V. O'Neill, D.L. Deangelis, J.B. Waide and T.F.H. Allen. *Hierarchical Concept of Ecosystems*. Princeton University Press, Princeton (1986).
23. C.A.Ross. Paleogeography and provinciality. In: *Paleogeographic Provinces and Provinciality*, C.A. Ross (ed.), *Society of Economic Paleontologists and Mineralogists Special Publications*, **21**, 1-17 (1974).
24. B. Runnegar. The Permian of Gondwana. In: *Proc. 27th Int. Geol. Congr.*, **1**, 305-339 (1984).
25. K.P. Schmidt. Faunal realms, regions, and provinces. *Q. Rev. Biol.* **29**, 322-331 (1954).
26. G.R.Shi. The Late Paiaeozoic brachiopod genus *Yakovlevia* Fredericks, 1925 and the *Yakovlevia transversa* Zone, northern Yukon Territory, Canada. *Proc. R. Soc. Vic.* **107**(1), 51-71 (1995).
27. G.R.Shi. Aspects of Permian marine biogeography: a review on nomenclature and evolutionary patterns, with particular reference to the Asian-western Pacific region. In: *Proc. Int. Conf. on the Permian System and Resources*. Balkema , Rotterdam (in press).
28. G.R. Shi and N.W. Archbold. Distribution of Asselian to Tastubian (Early Permian) circum-Pacific brachiopod faunas. *Mem. Ass. Australs. Palaeontols*, **15**, 343-351 (1993).
29. G.R. Shi and N.W. Archbold. A quantitative analysis on the distribution of Baigendzhinian-Early Kungurian (Early Permian) brachiopod faunas in the western Pacific region. *J. Southeast Asian Earth Sci.*, **11**, 189-205 (1995).
30. G.R. Shi, N.W. Archbold and L.P. Zhan. Distribution and characteristics of mixed (transitional) mid-Permian (Late Artinskian-Ufimian) marine faunas in Asia and their palaeogeographical implications. *Palaeogeogr., Palaeoclimat., Palaeoecol.* **115**, 241-271 (1995).
31. G.R. Shi, T.A. Grunt, N.W. Archbold and V.I. Manakov. Bipolar distribution of Permian Brachiopoda: implication for Permian correlation and palaeogeography. In: *30th Int. Geol. Congr. Abstracts*, **2**, 98 (1996).
32. G.R. Shi and L.P. Zhan. A mixed mid-Permian marine fauna from Yanji area, northeastern China: a paleobiogeographical reinterpretation. *The Island Arc* (1996, in press).
33. G.A. Thomas. Oldhaminid brachiopods in the Permian of northern Australia. *J. Palaeont. Soc. India*, **2**, 174-182 (1957).
34. V.I. Ustritskiy Permian climate. In: *Permian and Triassic Systems and Their Mutual Boundary*, A. Logan and L.V. Hills eds. *Can. Soc. Petroleum Geol. Mem.*, **2**, 733-744.
35. J. Utting (ed.). *Permophiles. A Newsletter of SCPS.* **28** (1996).
36. J.W. Valentine and C.L. May. Hierarchies in biology and paleontology. *Paleobiol.* **22**(1), 23-33 (1996).
37. R.Van der Voo. Paleomagnetism of North America: a brief review. In: Paleoreconstruction of the Continents, M.W. McElhinny and D.A. Valencio. *Geodynamics Series*, **2**, 159-176 (1981).
38. N.A. Yasamanov. Temperatures of Devonian, Carboniferous, and Permian seas in Transcaucasia and the Ural region. *Internat. Geol. Rev.*, **23**(9), 1099-1104 (1981).
39. A.M. Ziegler, M.L. Hulver and D.B. Rowley. Permian world topography and climate. In: *Glacial and Post-glacial Environmental Changes: Pleistocene, Permo-Carboniferous, Proterozoic*, I.P. Martini (ed.). Oxford University Press, Oxford (1996, in press).

Proc. 30ᵗʰ Int'l. Geol. Congr., Vol. 12, pp. 18-28
Jin and Dineley (Eds)

Ammonoid Palaeobiogeography of the South Kitakami Palaeoland and Palaeogeography of Eastern Asia during Permian to Triassic Time

MASAYUKI EHIRO

Institute of Geology and Paleontology, Tohoku University, Sendai 980-77, Japan

Abstract

Biogeographic analysis of Permian-Triassic ammonoids in eastern Asia suggests that the South Kitakami Paleoland, which remains as the South Kitakami Belt and Hayachine Tectonic Belt of Northeast Japan, was located in equatorial Tethys, with the South China continent and Khanka microcontinent closely situated, during the Middle to Late Permian. This proximity, especially between the South Kitakami and Khanka, persisted at least until Middle Triassic time.

Keywords: ammonoid paleobiogeography, Permian, Triassic, South Kitakami, eastern Asia

INTRODUCTION

The northeastern margin of the Asian continent and accompanying island arcs are composed largely of Late Paleozoic to Mesozoic accretion complexes. They are associated with some small continental blocks having pre-Mesozoic basements, such as the Bureya and Khanka in northeast China-Russian Far East and the South Kitakami in northeast Honshu, Japan (Fig.1). During Permian to Triassic time, the eastern part of the Tethys Sea was dotted with some large continental blocks: Sino-Korea, South China (Yangtze and Cathaysia) and Indochina blocks. The above mentioned small continental blocks might be situated near these large continents or might be parts of them. With regard to the location of these fragmentary continents during Permian to Triassic time, some different opinions have been presented as discussed later.

The South Kitakami Paleoland (SKP) is a land originally formed in a subduction zone along the northern margin of the Gondwanaland during the Early Paleozoic [11]. It remains as the South Kitakami Belt and Hayachine Tectonic Belt of Northeast Japan and consists of Caledonian basements and overlying shallow marine sedimentary rocks ranging in age from Silurian to Cretaceous. The paleogeography of the SKP during the Middle Paleozoic to Early Mesozoic has been studied mainly by using paleobiogeographic data, because no Paleozoic paleomagnetic data are available. In this paper the Permian-Triassic paleogeography of the SKP is discussed with relation to the Bureya continent and Khanka microcontinent, based on the ammonoid paleobiogeography.

PERMIAN AMMONOID PALEOBIOGEOGRAPHY IN EASTERN TETHYS

During Permian time, four ammonoid provinces, namely the Boreal, Equatorial American, Equatorial Tethyan and Peri-Gondwanan Provinces, are recognizable [9]. The Boreal Province (BP) is composed mainly of such regions as Urals, Novaya Zemlya, Siberia, Arctic Canada and Greenland; the Equatorial American Province (EAP) Texas, Coahuila, Guatemala and Columbia; the Equatorial Tethyan Province (ETP) South China, Southeast Asia, Iran, Tunisia and Sicily; and the Peri-Gondwanan Province (PGP) Australia, Himalayas, Salt Range and Madagascar. The Equatorial American and Equatorial Tethyan Provinces contain somewhat similar faunas, and so do the Boreal and Peri-Gondwanan Provinces.

Among the Early Permian (Asselian-Cathedralian) ammonoids, *Juresanites*, *Paragastrioceras* and *Uraloceras* characterize the BP and PGP, whereas genera belonging to the Perrinitidae, except for *Perrinites* reported from the Yukon Territory, Canada [29], are restricted to the EAP and ETP [31, 14] (Figure 2).

In eastern Asia, *Uraloceras* was reported from the western Inner Mongolia and northwestern Gansu in North China [23], and *Paragastrioceras* from the western Inner Mongolia [43]. At least one locality, yielding *Uraloceras*, in Gansu (loc.10 of fig.1 in Liang [23]) is considered to be located at or along the northeastern margin of the Tarim paleoland. Therefore, it is considered that the Tarim block and probably the northwestern margin of Sino-Korea block were parts of the BP in Early Permian time. On the other hand, the Lower Permian distributed in localities belonging to the ETP, such as South China [42], Thailand [15] and Timor [35], yield perrinitids. In Japan, *Properrinites* and *Paraperrinites* were reported from limestone in the accretion complexes of Southwest Japan: the Akiyoshi Limestone and the Permian of the Mino

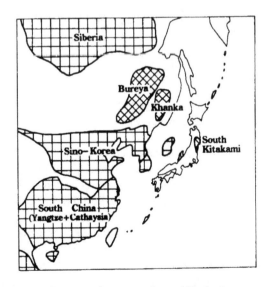

Figure 1. Index map of some continental blocks in eastern Asia.

Belt [32]. These limestone bodies might be formed in the ETP in Early Permian time. From the Lower Permian of SKP, on the other hand, only two cosmopolitan genera of ammonoids, *Agathiceras* and *Artinskia*, were reported [8].

In the early Middle Permian (Roadian), *Sverdrupites* and *Daubichites* characterized the AP [30], although *Daubichites* has also been reported from all other provinces [28]. There were, however, no genera endemic to the BP and PGP during the middle Middle Permian (Wordian) to Late Permian. On the contrary, the EAP and ETP had distinctive faunas during the Middle Permian (Fig.2). All the genera belonging to the Kufengoceratinae (Cyclolobidae) were restricted in the EAP and ETP, and have so far been reported from Texas, Coahuila and South China. Genera of the Paraceltitidae such as *Paraceltites, Cibolites, Nielsenoceras* and *Doulingoceras* occur from Texas, Coahuila, Tunisia and South China. The Cycolobid genus *Timorites* was also restricted to the EAP and ETP [9], occurring in such localities as Texas, Coahuila, Dzhulfa, Abadeh, Timor and South China, except for one locality in the Central Himalayas.

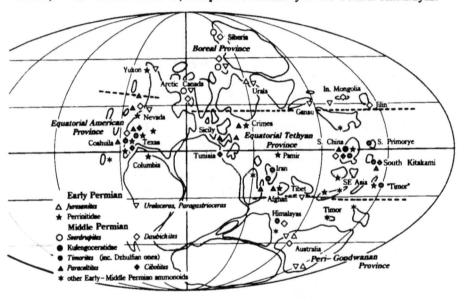

Figure 2. Geographic distribution of some selected Early-Middle Permian ammonoid genera (modified from Ehiro [9]). Permian ammonoid provinces are also shown.

Thirteen genera of ammonoids are known from the Middle Permian Kanokuran Series in the SKP [9, 10]. They include such genera characteristic of the ETP as *Paraceltites, Cibolites* and *Timorites* (Table 1). Recently, *Cibolites* was also collected from the Kurosegawa Tectonic Belt of Southwest Japan [21]. The Middle Permian formation of Southern Primorye, which is considered to have deposited at the southeastern margin of the Khanka microcontinent, also yields *Timorites* [22]. Thus the SKP, Kurosegawa Tectonic Belt and Southern Primorye (Khanka) were located in the ETP during the Middle Permian. Early Middle Permian strata in Jilin, which are considered to have deposited on the Bureya continent, yield 17 genera of ammonoids [23, 24], including *Waagenoceras* and *Daubichites*, but have no genera characteristic of the ETP.

Table 1. Distribution of some Middle Permian (Guadalupian) ammonoid genera in selected localities. Triangles are ones unknown from the South Kitakami Paleoland.

	1	2	3	4	5	6	7	8	9	10	11	
Agathiceras	O	O	O	O		O	O	O				
Pseudagathiceras						O	O					1 Australia
Sverdrupites										△	△	2 Himalayas
Daubichites	△	△		△			△				△	3 Timor
Roadoceras				O		O	O	O				4 South China
Pseudogastrioceras				O		O						5 Southern Primorye
Stacheoceras		O	O	O	O	O	O	O	O			6 South Kitakami
Kufengoceratid				O			O					7 Texas–Coahuila
Waagenoceras			O	O		O	O	O	O			8 Jilin–Inner Mongolia
Timorites			O	O	O	O	O					9 Gansu
Jilingites						O		O				10 Siberia
Propinacoceras	O		O	O		O	O	O	O			11 Arctic Canada
Medlicottia			O			O	O				O	
Eumedlicottia			O		O	O	O					
Paraceltites				O		O	O					
Cibolites				O		O	O					

Late Permian (Wuchiapingian-Changhsingian) xenodiscaceans, except for *Xenodiscus* (Xenodiscidae) and *Paratirolites* (Dzhulfitidae), and all Late Permian otocerataceans were restricted to the EAP and ETP, appearing in such localities as Coahuila, Dzhulfa, Abadeh and South China (Fig.3). *Pseudogastrioceras* has also been reported from various localities in the ETP, such as Dzhulfa, Abadeh, Southeast Asia and South China.

The Upper Permian Toyoman Series in the SKP yields 11 genera of ammonoids including *Timorites, Pseudogastrioceras* as some araxoceratids (otocerataceans) such as *Araxoceras, Prototoceras* and *Eusanyangites* [9] (Table 2). The Upper Permian of Southern Primorye also yields an araxoceratid *Eusanyangites* and xenodiscids such as *Xenodiscus* and *Iranites*? [40]. Therefore, the SKP and Southern Primorye are considered to have belonged to the ETP during the Late Permian as well as the Middle Permian.

TRIASSIC AMMONOIDS OF THE SOUTH KITAKAMI PALEOLAND AND ITS PALEOBIOGEOGRAPHIC SIGNIFICANCE

Tozer [37] distinguished the Triassic ammonoid biogeography into the Arctic, Pacific (East Pacific and West Pacific), Tethys, Germanic and Sephardic Provinces, although the boundaries between these provinces, especially between the West Pacific and Tethys, are unclear. There are many Triassic ammonoid genera having wide geographic distributions. In this paper, ammonoids occurring in both Pacific and Tethys Provinces are called the Pacific-Tethys type, and ones common in the Arctic, Pacific and Tethys Provinces cosmopolitan.

Triassic ammonoids have mainly been reported from the Spathian, Anisian and Lower Ladinian strata in the SKP. The Spathian ammonoid fauna of the Osawa Formation in the SKP [5, 4, 7, 16] is characterized by *Subcolumbites* and *Columbites* and dominated

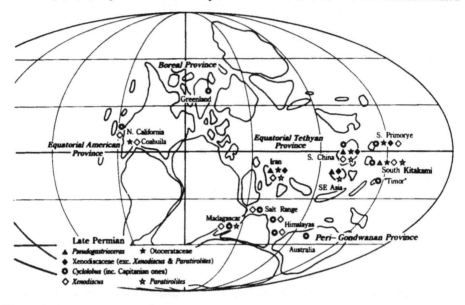

Figure 3. Geographic distribution of some selected Late Permian ammonoid genera (modified from Ehiro [9])

Table 2. Distribution of some Upper Permian (Wuchiapingian- Changhsingian) ammonoid genera in selected localities

	1	2	3	4	5	6	7	8
Pseudogastrioceras			O		O		O	
Stacheoceras		O	O		O	O		
Timorites					O	O	O	
Cyclolobus	O	O		O	O			O
Eumedlicottia		O		O	O		O	O
Neogeoceras				O	O			
Araxoceras			O		O		O	
Prototoceras					O		O	
Eusanyangites				O	O			
Xenodiscus		O	O	O	O	O	O	
Paratirolites		O	O		O		O	

1 Australia
2 Himalayas—Madagascar
3 South China
4 Southern Primorye
5 South Kitakami
6 Coahuila
7 Dzhulfa—Abadeh
8 Greenland

by the Pacific-Tethys type, with no Arctic ones such as *Arcttirolites, Arctmeekoceras, Boreomeekoceras, Sibirites* and *Parasibirites* (Table 3). Among 12 genera reported from the Osawa Formation, 19 genera, except for endemic *Eosturia*, have also been reported from many localities in the Tethys Province, and 17 genera from the Pacific, whereas ones from the Arctic are only 6. Sixteen genera known from the Osawa Formation are common to those from South China, and 15 to those from Southern Primorye. Thus the Spathian ammonoid fauna of the SKP is allied to those of South China and Southern Primorye, suggesting close geographic relationships (Fig. 4), although taxonomic positions of some Early to Middle Triassic ammonoids of Southern Primorye are in the course of revision [39].

The Anisian-Lower Ladinian strata (Fukkoshi, Isatomae and Rifu Formation) in the SKP yield such ammonoids as *Balatonites, Hollandites, Paraceratites* and *Protrachyceras*. Twenty-seven genera reported from these formations [33, 3, 6] are mostly of Pacific-Tethys type and there are no Arctic ones (Table 4). Twenty-three genera among them are also known from various localities in the Tethys Province, 20 from the Pacific and 8 from the Arctic. Among 21 genera of ammonoids from the Middle Triassic strata of Primorye [20, 38], 14 are common to those of South Kitakami. Therefore, it is considered that the SKP and Southern Primorye had close faunal relationships during the Middle Triassic as well as Early Triassic. On the other hand, the SKP has only six Middle Triassic ammonoid genera common to the faunas of South China and Siberia.

Table 3. Distribution of some latest Early Triassic (Spathian) ammonoid genera in selected localities. Triangles are ones unknown from the South Kitakami Paleoland.

	1	2	3	4	5	6	7	
Xenoceltites			O	O	O	O	O	
Arctotirolites						△		1 Iran
Nordophiceras			O	O		O	O	2 Timor
Arctomeekoceras						△		3 South China
Boreomeekocer;						△		4 South Kitakami
Dalmatites			O	O			O	5 Southern Primorye
Preflorianites			O	O		O	O	6 Siberia
Isculitoides	O	O	O	O	O	O	O	7 Nevada
Prohungarites		O		?			O	
Pseudosageceras	O	O	O	O	O	O	O	
Dinarites			O	O	O			
Stacheites	O		O	O			O	
Metadagnoceras	O	O	O	O	O		O	
Columbites	O		O	O	O		O	
Subcolumbites	O		O	O	O			
*Paragoceras**	O		O	O	O		O	*=*Arnautoceltites*
Prenkites	O	O	O	O	O			
Procarnites	O	O	O	O	O			
Sibirites						△		
Parasibirites						△		
Keyseringites				O	O	O	O	
Eosturia				O				
Danubites				O	O			
Eophyllites	O	O	O	O	O			
Leiophyllites	O		O	O	O		O	

In Upper Triassic strata of the SKP, ammonoids are very rare and all are cosmopolitan. Besides the SKP, some formations distributed in Southwest Japan yield Triassic ammonoids. They are Iwai Formation, Taho Limestone and Kamura Formation of Early Triassic, Zohoin Group of Middle Triassic, and Nabae Group, Kochigatani Group, Tanoura Formation and the Nakijin Formation of Late Triassic. Ammonoids from these Triassic strata are cosmopolitan, Pacific-Tethys type or Tethys type. No ammonoids closely related with the Arctic are known from Japanese Triassic.

Table 4. Distribution of some Middle Triassic (Anisian- Ladinian) ammonoid genera in selected localities. Triangles are ones unknown from the South Kitakami Paleoland.

	1	2	3	4	5	6	7	
Cuccoceras			O	O			O	
Balatonites			O	O	O		O	1 Iran
Hollandites	O			O	O		O	2 Timor
Beyrichites				O				3 South China
Gymnotoceras				O	O	O	O	4 South Kitakami
Frechites				O		O	O	5 Southern Primorye
Paraceratites	O		O	O	O		O	6 Siberia
Parakellnerites				O				7 Nevada– Idaho
Kellnerites				O				
Nevadites				O			O	
Hungarites	O			O	O			
Japonites		O	O	O	O			
Sturia	O	O		O	O			
Discoptychites			O	O	O		?	
Gymnites	O	O	O	O			O	
Anagymnites				O	O	O		
Kiparisovia						Δ		
Inaigymnites				O				
Epigymnites				O				
Ptychites	O	O		O	O	O	O	
Flexoptychites				O	O			
Danubites				O		O		
Stannakhites						Δ		
Karangatites						Δ		
Protrachyceras		O	O	O	O		O	
Arpadites	O			?				
Leiophyllites	O			O	O		?	
Ussurites	O			O	O	O		
Monophyllites	O	O		O			?	

PERMIAN-TRIASSIC PALEOGEOGRAPHY OF EASTERN ASIA BASED ON AMMONOID PALEOBIOGEOGRAPHY

There are two opinions concerning the paleogeographic situation of the SKP during the Permian. Based on the faunal similarity of corals [25] and bivalves [12, 13, 27], the SKP is considered to have been situated close to South China or Indochina. Kawamura and Machiyama [18] also placed the SKP at the southeast margin of South China near Indochina, because the tropical type coral reef bodies, mainly developed around the South China and Indochina continents during the Middle Permian, are known from the Middle Permian of the SKP. On the other hand, the Middle Permian brachiopod fauna of the SKP is closely allied to those of the Inner Mongolia and Jilin provinces of Northeast China [26, 36]. Opinions vary on the fusulinid and land plant biogeography. According to Ozawa [34], the Middle and Late Permian fusulinacean faunas of the SKP show striking similarities to those of Southern Primorye, South China and Southeast Asia, whereas Ishii [17] concluded that the South Kitakami belonged to the same fusulinacean territory as that of the Inner Mongolia-Jilin Province during the Middle Carboniferous to early Middle Permian. Lower Permian formations of the SKP contain plant fossils belonging to the Cathaysia flora [1, 2]. The Cathaysia floral province is divided into the North China and South China subprovinces. Asama [1] compared the Kitakami flora to the Shansi flora in North China, which belongs to the North China Subprovince, while Kimura [19] stated that it is difficult to determine whether it belongs to the North China Subprovince or South China one at the present state of knowledge.

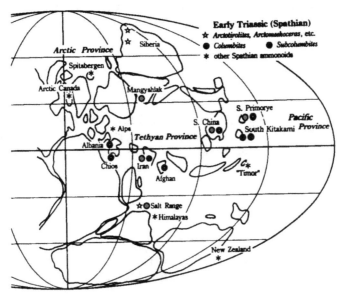

Figure 4. Geographic distribution of some selected late Early Triassic ammonoid genera in the Arctic, Tethyan and Pacific Provinces.

Permian ammonoid paleobiogeography appears analog to fossil corals and bivalves. Because, the Middle and Late Permian ammonoid faunas of the SKP are both closely related with those of South China belonging to the ETP. The Middle and Late Permian ammonoid faunas of Southern Primorye (Khanka microcopntinent) are also similar to them. Therefore, it is considered that these three continents were located closely in equatorial Tethys during Permian time (Figs. 2, 3). On the other hand, *Daubichites*- and *Waagenoceras*-bearing formations in Jilin (Bureya continent) and Inner Mongolia [23, 24, 43] differ from those of South China and SKP in that they have no Middle Permian paraceltitids, which characterize the ETP. The Bureya continent, belonging to the BP, may have been located far from the South China and SKP.

Concerning the Triassic biogeography, Nakazawa [27] considered that the molluscan fauna of the SKP had some relationship with that of Siberia, because a Middle Triassic bivalve, *Daonella densisulcata*, and some Middle Triassic ammonoid species, belonging to the genera *Hollandites*, *Ussurites*, *Sturia* and *Balatonites*, of the SKP are common to those of Siberia. These ammonoid genera are, however, all cosmopolitan, and Early-Middle Triassic ammonoid faunas of the SKP are lacking in the Arctic type ammonoids and consist mainly of Pacific-Tethys type as mentioned above. The ammonoid fauna implies that the SKP belonged to the Pacific or Tethys Province, not the Arctic, during the Early to Middle Triassic, leading to the conclusion that the SKP was situated at lower latitudes, with close geographic relationship among South China, SKP and Khanka, especially between the latter two, at least until Middle Triassic time (Fig.4).

The above mentioned paleogeographic reconstruction of the SKP and Khanka is consistent with the results of paleomagnetic study [41] which suggest that the paleolatitude of Southern Primorye was $1.0°N$ to $16.7°N$ during the Permian and $24.2°N$ at the Early Triassic.

Acknowledgments

I thank Professor Kunihiro Ishizaki of Tohoku University , Professor Brian F. Glenister of the University of Iowa, for their critical reading of the manuscript. This research was partly supported by a Grant-in-Aid for Scientific Research (C: 07640593) from the Ministry of Education, Science and Culture of Japan.

REFERENCES

1. K. Asama. Permian plants from Maiya in Japan. I. *Cathaysiopteris* and *Psygmophyllum*. *Bull. Nat. Sci. Mus., Tokyo* **10**, 139-153 (1967).
2. K. Asama and M. Murata. Permian plants from Setamai, Japan. *Bull. Nat. Sci. Mus., Tokyo* **17**, 251-256 (1974).
3. Y. Bando. The Triassic stratigraphy and ammonite fauna of Japan. *Sci. Rep. Tohoku Univ., 2nd Ser.* **36**, 1-137 (1964).
4. Y. Bando and M. Ehiro. On some Lower Triassic ammonites from the Osawa Formation at Asadanuki, Towa-cho, Tome-gun, Miyagi Prefecture, Northeast Japan. *Trans. Proc. Palaeont. Soc. Japan, N.S.* **127**, 375-385 (1982).
5. Y. Bando and S. Shimoyama. Late Scythian ammonoids from the Kitakami Massif. *Trans. Proc. Palaeont. Soc. Japan, N.S.* **94**, 293-312 (1974).
6. M. Ehiro. A new species of *Parakellnerites* (Triassic ammonoid) from the Rifu Formation, Northeast Japan. *Saito Ho- on Kai Mus. Res. Bull.* **60**, 1-6 (1992).
7. M. Ehiro. Spathian ammonoids *Metadagnoceras* and *Keyserlingites* from the Osawa Formation in the Southern Kitakami Massif, Northeast Japan. *Trans. Proc. Palaeont. Soc. Japan, N.S.* **171**, 229-236 (1993).
8. M. Ehiro. Cephalopod fauna fo the Nakadaira Formation (Lower Permian) in the Southern Kitakami Massif, Northeast Japan. *Trans. Proc. Palaeont. Soc. Japan, N.S.* **79**, 184-192 (1995).
9. M. Ehiro. Permian ammonoid fauna of the Kitakami Massif, Northeast Japan - Biostratigraphy and paleobiogeography-. *Palaeoworld*, **9** (*Proc. ISP'94, Guiyang)* (1997, in press).
10. M. Ehiro. Permian cephalopods of Kurosawa, Kesennuma City, Southern Kitakami Massif, Northeast Japan. *Palaeont. Res.* **1** (1997, submitted).

11. M. Ehiro and S. Kanisawa. Formation and evolution of the South Kitakami Paleoland during the Paleozoic, with special reference to the geodynamics of eastern Asia (IGCP-321). *Japan contribution to the IGCP, 1996: Geologic development of the Asian-Pacific region, with implications in the evolution of Gondwanaland*, IGCP Nat. Comm. Japan (Ed.). pp.43-50. IGCP Nat. Comm. Japan, Shizuoka (1996).

12. Z. Fang. A preliminary study of the Cathaysia faunal province. *Acta. paleont. Sinica* 24, 344-349 (1985). [in Chinese with English abstract]

13. Z. Fang and D. Yin. Discovery of fossil bivalves from Early Permian of Dongfang, Hainan Island with a review on glaciomarine origin of Nanlong diamictites. *Acta Palaeont. Sinica* 34, 301-315 (1995). [in Chinese with English abstract]

14. W.M. Furnish. Permian stage names. *The Permian and Triassic Systems and their mutual boundary*. A. Logan and L.V. Hills (Eds.). pp.522-548. Canad. Soc. Petrol. Geologists, Mem. 2, Calgary (1973).

15. B.F. Glenister, W.M. Furnish, Z. Zhou and M. Polahan. Ammonoid cephalopods from the Lower Permian of Thailand. *Jour. Paleont.* 64, 479-480 (1990).

16. T. Ishibashi, K. Hasegawa, Y. Sato, K. Kamada and M. Murata. Ammonoids from the Inai Group in the Southern Kitakami Massif - Part 1. Osawa Formation (Lower Triassic)-. *Abst. 143rd Reg. Meet. Palaeont. Soc. Japan*, 37 (1994).

17. K. Ishii. Comparison of fusulinacean faunas between Inner Mongolia-Jilin Province and Japan. *Pre-Jurassic geology of inner Mongolia, China. Report of China-Japan Cooperative Research Group, 1987-1989*. K. Ishii, X. Liu, K. Ichikawa and B. Huang (Eds.). pp.189-199. Osaka City University, Osaka (1991).

18. T. Kawamura and H. Machiyama. A Late Permian coral reef complex, South Kitakami Terrane, Japan. *Sed. Geol.* 99, 135-150 (1995).

19. T. Kimura. Geographical distribution of Palaeozoic and Mesozoic plants in East and Southeast Asia. *Historical biogeography and plate lectonic evolution of Japan and Eastern Asia*. A. Taira and M. Tashiro (Eds.). pp. 135-200. Terra Scientific Publishing Company, Tokyo (1987).

20. L.D. Kiparisova. Paleontological basis of the stratigraphy of the Triassic formations in Ussuriland. Part 1. Cephalopods. *Trans. All-Union Geol. Res. Instit., N.S.* 48, 1-278 (1961). [in Russian]

21. H. Koizumi, K. Mimoto and N. Yoshihara. Permian ammonoid from Katsura, Sakawa, Kouchi Prefecture, Southwest Japan. *Chigaku Kenkyu* 43, 29-33 (1994). [in Japanese]

22. G.V. Kotlyar, Y.D. Zakharov, G.S. Kropatcheva, G.P. Pronina, I.O. Chedija and V.I. Burago. *Evolution of the latest Permian biota. Midian regional stage in the USSR*. Academy of Sciences, USSR, Leningrad (1989). [in Russian]

23. X. Liang. Early Permian cephalopods from northwestern Gansu and western Nei Monggol. *Acta Palaeon. Sinica* 20, 485-500 (1981). [in Chinese with English abstract]

24. X. Liang. Some Early Permian ammonoids from Jilin and Nei Monggol. *Acta. Palaeont. Sinica* 21, 645-657 (1982). [in Chinese with English abstract]

25. M. Minato and M. Kato. Waagenophyllidae. *Jour. Fac. Sci. Hokkaido Univ., Ser. IV* 12, 1-241 (1965).

26. K. Nakamura and J. Tazawa. Faunal provinciality of the Permian brachipods in Japan. *Pre-Cretaceous terranes of Japan*. K. Ichikawa, S. Mizutani, I. Hara, S. Hada and A. Yao (Eds.). pp.313-320. Osaka City University, Osaka (1990).

27. K. Nakazawa. Mutual relation of Tethys and Japan during Permian and Triassic time viewed from bivalve fossils. *Shallow Tethys 3*. T. Kotaka, J.M. Dickins, K.G. McKenzie, K. Mori, K. Ogasawara and G.D. Stanley (Eds.). pp.3-20. Saito Ho-on Kai, Sendai (1991).

28. W.W. Nassichuk. Permian ammonoids from Devon and Melville Islands, Canadian Arctic Archipelago. *Jour. Paleont.* 44, 77-97 (1970).

29. W.W. Nassichuk. Permian ammonoids and nautiloids, southwestern Eagle Plane, Yukon Territory. *Jour. Paleont.* 45, 1001-1021 (1971).

30. W.W. Nassichuk. Permian ammonoids in the Arctic regions of the world. *The Permian of Northern Pangea, Volume 1: Paleogeography, paleoclimates, stratigraphy.* P.A. Scholle, T.M. Peryt and D.S. Ulmer-Scholle (eds.). pp.210-235. Springer-Verlag, berlin 91995).

31. W.W. Nassichuk, W.M. Furnish and B.F. Glenister. The Permian ammonoids of Arctic Canada. *Geol. Surv. Canada, Bull.* **131**, 1-56 (1965).

32. T. Nishida and Y. Kyuma. Ammonoid faunas of the Upper Carboniferous and Lower Permian of Japan. *Newslet. Co-op. Res. Gr. C/P Boundary* **1**, 123-128 (1991).

33. Y. Onuki and Y. Bando. On the Inai Group of the Lower and Middle Triassic System. *Contr. Inst. Geol. Paleont., Tohoku Univ.* **50**, 1-69 (1959). [in Japanese with English abstract]

34. T. Ozawa. Permian fusulinacean biogeographic provinces in Asia and their tectonic implications. *Historical biogeography and plate tectonic evolution of Japan and Eastern Asia.* A. Taira and M. Tashiro (eds.). pp. 45-63. Terra Scientific Publishing Company, Tokyo (1987).

35. J. P. Smith. Permian ammonoids of Timor. *Mijn. Nederland.-Ind., Jaarb., Jaarg. (1926)* **1**, 1-91 (1927).

36. J. Tazawa. Middle Permian brachiopod biogeography of Japan and adjacent regions in East Asia. *Pre-Jurassic geology of inner Mongolia, China. Report of China-Japan Cooperative REsearch Group, 1987-1989.* K. Ishii, X. Liu, K. Ichikawa and B. Huang (eds.). pp.213-230. Osaka City University, Osaka (1991).

37. E. T. Tozer. Triassic Ammonoidea: Geographic and stratigraphic distribution. *The Ammonoidea: the evolution, dassification, mode of life and geological usefulness of a major fossil group.* M.R. House and J. R. Senior (eds.). pp.397-431. Academic Press, London (1981).

38. Y.D. Zakharov. *Lower Triassic ammonoids of East SSSR.* Academy of Sciences, USSR, Moskva (1978). [in Russian]

39. Y.D. Zakharov. (1996, pers. comm.)

40. Y.D. Zakharov and A.M. Pavlov. Permian cephalopods of Primorye region and the problem of Permian zonal stratigication in Tethys. *Correlation of Permo-Triassic sediments of East USSR.* Y.D. Zakharov and Y.I. Onoprienko (eds.). pp.5-32. Far Eastern Science Centre, Academy of Sciences, USSR, Vladivostok (1986). [in Russian]

41. Y.D. Zakharov and A.N. Sokarev. Permian-Triassic paleomagnetism of Eurasia. *Shallow Tethys 3*, edited by T. Kotaka, J.M. Dickins, K.G. McKenzie, K.Mori, K. Ogasawara and G.D. Stanley, Jr. (eds.). pp.313-323. Saito Ho-on Kai, Sendai (1991).

42. Z. Zhou. First discovery of Asselian ammonoid fauna in China. *Acta Palaeont. Sinica* **26**, 130-148 (1978). [in Chinese with English abstract]

43. G. Zhu and H. Sheng. Early Permian ammonoids from Hunggermiao of Abga Qi, Inner Mongolia. *Prof. Pap. Stratigr. Palaeont.* **21**, 65-78 (1988). [in Chinese with English abstract]

Proc. 30ᵗʰ Int'l. Geol. Congr., Vol. 12, pp. 29-53
Jin and Dineley (Eds)
© VSP 1997

Palaeobiogeographic Evolution of Permian Brachiopods

JIN YU-GAN SHANG QING-HUA
Nanjing Institute of Geology & Palaeontology, Chinese Academy of Sciences, Nanjing, 210008 China

Abstract

Palaeobiogeographic evolution of Permian brachiopods was investigated based on a quantitative analysis proceeded from the data base which collected 28,000 records of brachiopod taxa from 450 stratigraphic levels of 289 localities all over the world. The Otsuka coefficient was selected as the best quantitative indices in revealing faunal affiliation. The biogeographic affinities of coeval brachiopod faunas can be evidently revealed by the results of cluster analysis when the faunal analogs rooted in community relationship, bipolar distribution and low efficiency of collections have been identified.

Changes in faunal affinity demonstrate that biogeographic evolution of Permian brachiopods had been profoundly influenced by the consolidation of the Pangea, which was finalized by the close of Ural seaway at the end of Artinskian Stage, and led to a sharp differentiation of a pro-Pangea faunas and the Tethyan faunas. The brachiopod faunas in peri-Pangea shelves are dominated by a group of such forms as *Spiriferella, Neospirifer,* the syringothyrids and the buxtoniids, which directly descended from Cisuralian faunas. These faunas are conservative as they extended upward to the Lopingian with little evolutionary changes. The faunas dwelt in off-shore islands of Tethys and Panthalassa were characterized by newly developed endemic brachiopods but lacked pro-Pangea taxa. Between these two end-member faunas there were transitional faunas occurring in the peripheral islands of Pangea.

Keywords: Permian, Cisuralian, Kungurian, Guadalupian, Lopingian, brachiopods, palaeobiogeography, cluster analysis, Pangea

INTRODUCTION

As one of the dominant benthic fossil groups in the Permian, the brachiopods have been extensively documented both geographically and stratigraphically. Unlike the other main marine invertebrate groups known in considerable details such as the fusulinids, rugose corals, bryzoans and ammonoids, this group is abundant worldwide. Frequently, researchers have endeavored to recognize the affinities between brachiopod faunas of different regions. As a result more obvious affinities, such as those between the Tethys and the Boreal brachiopods, have been known for many years. Among those who have been especially active in the evaluation of Permian brachiopod biogeographic evolution, Stehli [26] and Ustrisky [32] have dealt primarily with the relation between faunal distribution and major climate belts, such as the cold, temperate and warm water faunas based on selected endemic taxa. In a series of papers, Stehli suggests that the distribution of Permian brachiopods does not support post-Permian continental displacement [28, 26]. This brachiopods-based suggestion has a profound influence and could even gain support from recent studies such as that on Permian bivalve biogeography [4].

Attempts to quantitatively measure brachiopod faunal affinities and differences have been made by Stehli [26], Grant and Cooper [3], Waterhouse and Bonham-Carter [34]. This approach employs high taxonomic categories, rather than genera or species to derive the model. For instance, based on the cluster analysis of a variety of 54 brachiopod families in 64 localities, Waterhouse and Bonham-Carter [34] examine the differentiation of three nearly independent realms during all three epochs of the Permian. However, the biogeographic similarities revealed at family level are commonly overemphasized and thus, further subdivision of biogeographic units is hardly possible. It has been found from past experiences that genera and subgenera provided the best taxonomic level for investigating faunal biogeographic affinities. More recently, interests in quantitative analysis of Permian brachiopods are revived with new focuses by Archbold and Shi [1]. Their interests include an extensive test on various methods of quantitative analysis [24], a thoroughly survey on the available records of Permian brachiopod genera and species with particular emphasis on the circa-Pacific regions [24, 25].

We propose to explore the palaeobiogeographic evolution of Permian brachiopods based on a relatively complete data base, which enables us to test the various existing patterns of Permian biogeography in terms of the faunas susceptible to statistical evaluation. The evaluation can be undertaken in several ways, such as tracing dispersal history of the endemic taxa or some particular assemblages, comparison of indices of similarities and differences among brachiopod faunas and percentages of common taxa between two faunas. This paper reports the results of a cluster analysis of similarity indices of paired brachiopod faunas.

BASIC DATA

Reference Sources
This study benefits greatly from a computerized list of papers on Permian brachiopods provided by Mr. Rex Doescher in Smithsonian International Brachiopod Information Center (SIBIC). It contains 1,915 reports published by 92 senior authors during the last 150 years. Supplemental data are from our own collections, particularly those unpublished, such as dissertations, files and reports for dormitory distribution. The fossils records were sorted in term of their reliability. The brachiopod faunas documented in descriptive papers are regarded as the most reliable. The lists of fossils given by brachiopod experts are ranked as the less reliable, which were only cited for the areas without the most reliable data. Totally, we have collected 10,118 records of species and 4,211 records for the families and genera. The brachiopod associations recorded are from 229 stratigraphic units of 124 localities in China and 271 stratigraphic units of 107 localities in other countries. Each locality usually represents a particular depositional basin rather than an outcrop or outcrop area.

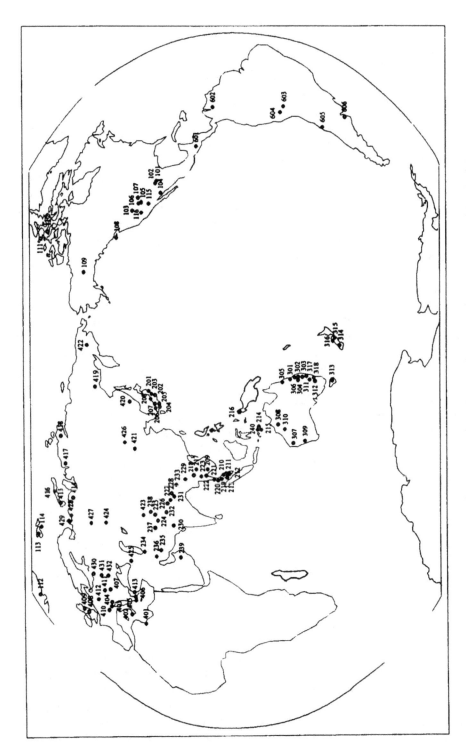

Figure 1. Location of Permian brachiopod faunas. Locality number in China are not shown in the map.

Adequacy of Sampling

In order to eliminate the bias caused by unequal collecting and disproportionate representation of different communities, available faunal data were sorted based on their adequacy. Assuming that a majority of families distributed worldwide can be found from an area collected extensively, Stehli and Grant [26] have suggested a collecting efficiency index, i.e., the percentage of cosmopolitan families for a particular locality. Obviously, this assumption only holds under the condition that a whole spectrum of brachiopod communities had been developed and preserved in the studying area. But in reality, only some areas satisfy such pre-requirements. A broad spectrum of environments from nearshore to open marine, clastic to carbonate-dominated facies is not always well developed or preserved in every area. Therefore, very often the brachiopods from an area may not contain most cosmopolitan families. In this study, we leave out under-sampled localities which comprise representatives of less than 3 cosmopolitan families. Usually, these localities appear in the inconsistent clusters with rather low similarity coefficient. Using too many of these localities will result in a messy dendrogram. Accordingly, we selected 96 localities for Cisuralian faunas, 60 for the Kungurian, 72 for the Guadalupian and 42 for the Lopingian.

Palaeogeographic Base Maps

Several reconstructions of Permian continents were developed during the last decade [37, 21, 18, 20]. Among them, those presented by Ross and Ross [20] are especially fascinating as it attempts to delineate the palaeogeographic pattern with emphasis on palaeobiogeographic evidences of well studied fossil groups of the Permian. The hypothetical distribution of microcontinents and arcs is interesting because it suggests the microcontinents and arcs with the Tethyan faunas as archipelagos scattered within Panthalassa rather than a complex of blocks essentially within the gulf-shaped ocean in other reconstruction maps. In this paper, we use the recently revised maps and their data sets provided by A. M. Ziegler and his collaborators [38]. The data sets readily allow us to plot the localities in the map based on their longitude and latitude. However, some parts need to be modified substantially in accordance with the provinciality of brachiopods. For instance, the Central Iran and Transcaucasus are placed in a latitudinal position close to South China. In the existing reconstruction maps, a vast Mongol-Okhatsk Ocean between the Mongol-Amur block and the Siberia continent is commonly accepted [15, 6]. The Mongol-Amur block is designed to contain most areas between North China and Siberia continent in the early Permian, and collided with North China in the late Permian. This arm-shaped block extended southeastward into the Panthalasia. Permian brachiopod faunas from the component areas are apparently different in biogeographic affinities [16].

Taxonomic and Stratigraphic Framework

To achieve a measure of uniformity in taxonomic treatment, the generic re-assignment of brachiopods was made from plates by Dr. Shen Shu-zhong from University of Mining and Technology of China and ourselves. Family assignment for the genera was made primarily in accordance with the classification of the revised edition of Part H of <the Treatise on Invertebrate Palaeontology>. Totally, 654 Permian named genera of brachiopod are used, including 390 genera of Cisuralian, 340 genera of Kungurian, 401 genera of Guadalupian and 237 genera of Lopingian.

The time framework used is the Permian chronostratigraphic scheme proposed by the Permian Subcommission, IUGS [10]. The Cisuralian Epoch is here divided into two independent units since its time span is much longer than the upper two epochs. Most brachiopod bearing beds of the Permian can be dated at the epoch level, but in some area more than the data warrant, and thus, will bias the result of our analysis.

The isotopic age for the boundaries of this system and its component series are adopted from Chuvashov et al. [2] for the Cisuralian Series, and from Roberts et al. [17] for the rest parts. Totally 127 fossil-bearing beds are compiled in a correlation chart of 107 areas or localities.

DATA ANALYSIS

Cluster Analysis

As a useful biogeographic analysis technique, cluster analysis is applied to the Permian brachiopod data in order to reveal biogeographic affinities of faunas. Using a program formulated by the junior author (SQH) based on Foxpro, a triangular matrix of numbers of common genera for paired faunas was produced. Then, we tested Jaccard's, DC and Otsuka Coefficient as a measure of similarity, and found that the Otsuka Coefficient represents the previously recognized biogeographic features of Permian brachiopods more closely than the other coefficients. For example, when the Otsuka coefficient is used as an indicator of clustering, the adequate samples of neighboring provinces would not scatter in clusters as disperse as they do when other coefficients are used.

We used the Q-mode for grouping clusters mainly because it is more straightforward in representing biogeographic relation between various areas. The triangular matrixes of similarity value between all possible pairs of faunas were calculated using a software named MVSP. Though there are two slight different methods of clustering based on progressive clumping of faunas with high mutual similarity, i.e., the weighted pair-group method and the unweighted pair-group method, the dendrogram generated by these two methods are essentially overlapping [34]. In this study we use the unweighted pair-group method.

Palaeobiogeographic Reality

Before attempting to identify the most reasonable biogeographic interpretation of the dendrograms, we have to consider some of the problems which inevitably become involved if we deal with the dendrogram of fossil data. The dichotomies of different association levels will be assembled by the similarities of both biogeographic and local ecological features. they also can be forced by method of cluster analysis itself on the faunas even when there is no distinct discontinuity of faunas of this realm. The clusters representing Tethyan faunas of the Lopingian serve as a good example for the gradient of faunal change (Fig. 9). The Lopingian brachiopods from western Guizhou, which have been fully collected and well described, form the core clusters of Tethyan faunas. The similarity coefficients decrease from more than 0.6 for the core clusters to less than

0.2 for the clusters of faunas from marginal areas of the Tethys. However, the transition between two endemic faunas is not always exhibited as an array of clusters with progressively variation of similarity coefficients. The cluster analysis forces all samples to cluster and therefore, may obscures the gradient of faunal change. Among Kungurian faunas from the areas bridging the Tethyan and Gondwana realms, those from Thailand, Kashmir, Shan State and Malasiya were grouped into the clusters of Tethyan faunas while those from Rutog, Dingri, Arunchal, Garhual, South Pamir, and Wardak of Afghanistan, into the Gondwana faunas. The association levels of similarity coefficient for these clusters are commonly low, though clusters with low coefficient seen in the connection branch may indicate lack of sufficient data. In both cases, the clusters are not very consistent.

Monograph Overprint

We believe that the similarity of Permian brachiopod faunas would be distorted by the over splitting taxonomic work for a few faunas. A pilot analysis on such faunas proved that introduction of numerous new genera for a brachiopod fauna would not result in the isolation of that area. A total of 36 new genera were named for the brachiopods from the Capitanian patch reefs in Lengwu of Zhejiang, South China [14]. This fauna appears in the cluster for the faunas of southeastern China since a substantial number of common generic and species names of Capitanian brachiopods in this region were also retained for the Lengwu Fauna in addition to new taxa. On the other hand, the Permian brachiopods of West Texas are distinctly separated from the others at high association levels, because it has been keeping a traditional nomenclature of fossils for more than a century.

Benthic Associations

As discussed in the part about efficiency of data collecting, the brachiopods from an area often do not comprise a full span of communities and therefore, the cluster analysis will be profoundly influenced by local ecological factors. Usually, the most diverse community is regarded as the diagnostic representative of regional faunas in recognizing plaeogeographic province. In cluster analysis, however, it can not be excluded that the similarity only represents most dominant communities, neither the most diverse community nor a span of communities from on-shore to off-shore. For example, Cisuralian brachiopods from nearly all areas of North China occur in marine beds of highly cyclic coal-bearing sequences and consequently, comprises near shore, clastic bottom communities with moderate diversity. High diverse association that has been recorded from a single of locality in North China was assembled into a separate cluster with other high diverse faunas.

Ecological influence are especially pronounced in the area with overlapping faunas of different provincial affinities such as the Kungurian and Guadalupian brachiopods in Nei Mongol-Amur province and the Kungurian brachiopods in Cimmerian block. The brachiopod community associated with clastic facies show strong faunal affinities of Boreal provinces while the community with massive carbonate rocks contain significant number of Tethyan elements [8, 9]. Nevertheless, as one of the most important features in defining paleobiogeographic units, the framework of brachiopod community remains to be reconstructed.

Bipolar Distribution

It has been documented for years that Permian brachiopods display a bipolar distribution [36], that is, the faunas from areas of high latitude share significant number of genera. These faunas may occur in the same cluster though they are not from the same hemisphere. To avoid such confusion, cluster analysis were applied to the faunas from southern hemisphere and those from northern hemisphere separately with the faunas from low latitudinal areas.

HISTORICAL BIOGEOGRAPHY

Cisuralian (296-280 Ma)

From the dendrogram for the faunas of southern hemisphere (Fig.3), three major clusters are distinct. They are corresponding to the Gondwanan (clusters C-F), the Grandian-Cordillera (cluster B), and the rest of pan-equatorial provinces (clusters H-N). The similarity coefficients between them are less than 0.2. Among other distinctive clusters, cluster A includes two faunas from Xizang of China of which both are poorly represented and thus, are not comparable with others; cluster G consists of three faunas from the late Artinskian beds in South China which are substantially younger in age than the others.

The Gondwanan provinces comprise the western Australian province with the similarity coefficient of more than 0.45 (cluster F), the Austrazean (Eastern Australian and New Zealand) province with the similarity coefficient of 0.3 (cluster E), and the faunas appeared in northwestern Gondwana extending from West Timor to Wardak of Afghanistan (cluster C and D). Some other faunas from northwestern Gondwana form a cluster (H) of the pan-equatorial group. It should be noted that the brachiopods from the Dorud Formation of Elburz Mt., North Iran are based on a preliminary report [7]. The brachiopods recently obtained from the same locality by us prove that such characteristic genera of Gondwanan Realm as *Taeniotharus* and *Stepanoviella* ought to be identified as *Juressania* and *Linoproductus*. Occurrence of abundant fusulinids of *Pseudoschwagerina* indicates a non-Gondwanan affinity as well.

Faunas referred to as a brachiopod community offshore to the *Eurydesma* fauna in the Gondwana shelves have not been included in the clusters because their diversity is rather low. These faunas are characterized by a dominance of such linoproductids as *Bandoproductus* and *Stepanoviella*. The brachiopods referred to "*Cancrinella fergalensis*" in Argentina and Australia appear to be identical to *Bandoproductus* in generic features. These faunas have been recorded from the *Cancrinella fergalensis* Zone in the Parana Basin of Argentina, the Callytharra Formation and its corresponding beds in western Australia, the Umaria Bed and its equivalents of India Peninsula, and the top part of the diamictics group nearby Lhasa of Xizang.

Among the clusters of the pan-equatorial provinces , the similarity coefficient for the faunas from the Tarim province (cluster K), the North China province (cluster J), the Central America province including Bolivia and Peru (cluster N) is more than 0.4. The cluster M contains rather dispersed localities with the similarity coefficient of more than 0.3. All of them are highly diverse faunas associated with carbonate facies of open marine. The southern Alps and Abadeh of Iran, the Carnic Alps and Zhen'an of Shaanxi are loosely lumped in the clusters I and L respectively, yet they can not be regarded as representatives of a single biogeographic province.

The dendrogram for the faunas from Northern Hemisphere (Fig. 3) includes clusters which have been revealed in the preceding dendrogram such as the Grandian-Cordillera provinces (cluster A), Artinskian association of South China (cluster B), the highly diverse faunas from pan-equatorial provinces (cluster G and H), North China (cluster I) and Tarim province (cluster J). In addition, there is a cluster representing Boreal provinces (cluster D), which can be further divided into two subclusters, the Siberian province consisting of faunas from Verkhoyan, Taymir and Novaya Island, and the Ural-Franklin province including the northern Urals, Kanin Acheaped of Russia and Vancouver of Canada (clusters D1 and D2). A cluster referable to the Beishan-Mongol province (cluster E) consists of the faunas from eastern Xinjiang, western Nei Mongol and the Urals. The clusters of highly diverse faunas include that from Wang Saphung of Thailand, Hiroshima and Shikogu of southern Japan (cluster G), and the faunas from eastern Nei Mongol and Jilin (cluster H). The widespread occurrences of the highly diverse faunas implies that all these regions, including Southern China, Northern China, Tarim, Nei Mongol and the Urals belong to a broad biogeographic realm covering the low latitudinal regions of Eurasia. In contrast, the Boreal province is not comparable with the Gondwana Realm in size. The biogeographic distribution of the fusulinids enables us to reach the same conclusion that the Ural-Franklin Region is indistinct from the Tethyan Realm [18].

It is noteworthy that the brachiopods from northwestern California falls in the cluster of Franklin-Ural province (cluster E) or in that of Cimmerian province (cluster H), but is not related to the clusters of Cordillera faunas. Other localities with low similarity coefficient are Southern Alps and Jiuquan of Beishan (cluster C), Ukraine, Ejin-Qi of Nei Mongol and Hebei of North China (cluster F), and so forth. Again, the grouping of these clusters shows an overlapping of faunal distribution or some gradual change from south to north.

The distribution of Cisuralian brachiopods follows well the paleolatitudianal gradient. The Andean, Austrazean, and the Perigondwana province of the Gondwana Realm distinctly differ from the others. A large group of clusters with gradual increasing similarity coefficient is prominent in the dendrogram for all Cisuralian brachiopod faunas. It suggests that the tropic and subtropic provinces were wider in latitude range than those in succeeding epochs. The scattering distribution of highly diverse brachiopod faunas further proves the vast extension of the tropic and subtropic faunas and that the cool-water faunas of the Boreal Realm was probably restricted to the Siberian province.

Kungurian (280 - 272 Ma)
The clusters of Kungurian brachiopods appears to be better defined and more consistent than those of Cisuralian brachiopods. Only few brachiopod faunas fall into the clusters with very low similarity coefficient for the reason of under- sampling. From

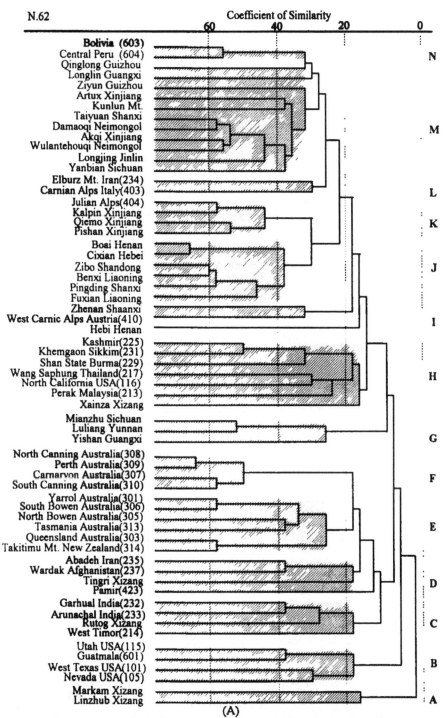

(A)

Fig 2. Dendrograms respectively for Cisuralian brachiopods from the Southern Hemisphere (A), and the Northern Hemisphere (B).

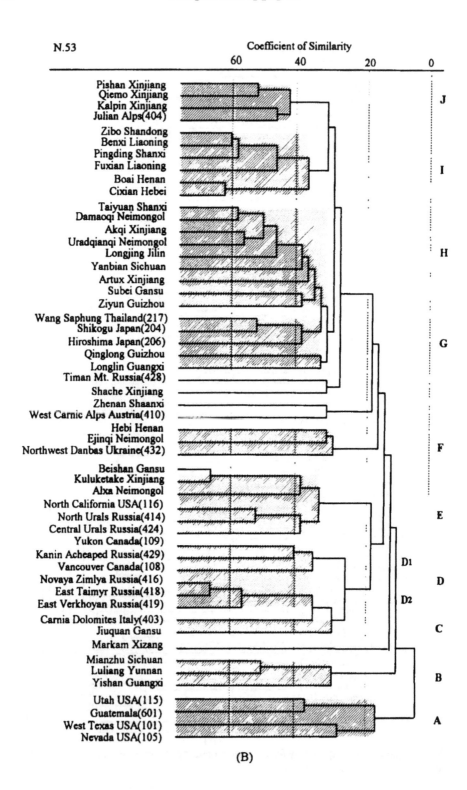

(B)

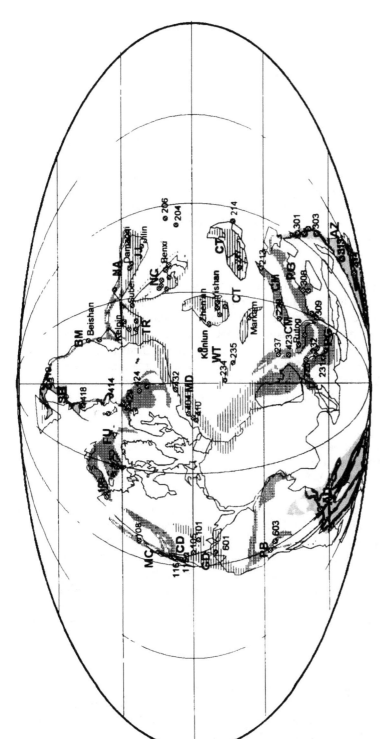

Figure 3. Palaeobiogeographic provinciality of Cisuralian brachiopods. Names of Permian marine biotic realms and provinces are adopted from Stehli [29], Yancey [35] and Ross [19]: AD-Andean province, AZ-Austrazean province, BM-Beishan-Mongol province, CM-Cimmerian province, CD-Cordillera province, CT-Cathaysia province, FU-Franklin-Uralian province, GD-Grandian province, JJ-Jilin-Japan province, MC-McCloud province, MD-Moldavian province, NC-North China province, NA-Nei Mongol-Amur province, PG-Peri-Gondwana province, PB-Peru-Bolivian province, SB-Siberian provinces, WT-West Tethys province. Polar provinces *dark shaded*, high latitudinal provinces *medium shaded*, and pan-equatorial provinces *light shaded*.

the dendrogram of Southern Hemisphere (Fig. 5B), there are clusters representing the Tarim province (cluster A), Cathaysia province (cluster B), the Grandian-Cordillera provinces (cluster C), the Western Australia (cluster D) and the Himalayan faunas (cluster F) of Peri-Gondwana province, the Cimmerian province (cluster E) and the Austrazean province (cluster G). Among the faunas from areas distant from the above mentioned provinces, that from Saudi Arabia is grouped together with the Himalayan faunas. It represents the western extension of peri-Gondwanan province. The central Chili fauna is grouped into the cluster consists of the faunas from Lhasa block including Xainza and Rutog of Xizang. This fact indicates that this fauna marks a South America province, which occupied a palaeolatitudinal position comparable to the Cimmerian province. The fauna from southern Pamir joins the cluster for the faunas from eastern Australia and New Zealand, and thus shows fairly strong affinities with the Gondwanan faunas. The brachiopods from northern California is correlated with the Sibumasu faunas (cluster E) with a similarity coefficient more than 0.3. The faunas from Abadeh of Iran, southern Qilian and western Cambodia are included in the group of clusters of Tethyan provinces (clusters A and B).

From the dendrogram of Northern Hemisphere (Fig. 5A), in addition to the clusters overlapped by those in the preceding dendrogram, there are clusters corresponding the Beishan-Mongolia province (cluster F1), Nei Mongol-Amur province (cluster F2), and the Franklin-Uralian province including the faunas from Spitsbergen, Arctic Canada, northern Urals (cluster E). The faunas from Verkhoyan and Omolon occur in the same cluster (D) that represents the Siberian province. Besides, the brachiopods from the terrigenous clastic beds in Nei Mongol are also grouped with the faunas of the Franklin-Uralian province, and the brachiopods from northern Japan arc tied to this Siberian cluster with low similarity coefficient.

Differing from the dendrogram of Cisuralian brachiopods, all faunas from Beishan, Nei Mongol and northern Japan show a close relation to the faunas of Boreal Realm rather than the Tethyan faunas as before. The general trend of brachiopod development from the Cisuralian to the Kungurian in these areas is marked by a significant increase in stocks that had been developed in Boreal regions. Kungurian brachiopods from Beishan and southern Mongolia are closely related to each other and also to those of the Franklin-Uralian province [12, 13]. In addition to the mingling of Boreal and Tethyan taxa, a unique span of communities of Cisuralian brachiopods occurred in Nei Mongol - Amur province. The brachiopod community in clastic facies shows strong faunal affinities of Boreal provinces while that from the carbonate bank facies contains more Tethyan elements. The mingling aspects have led some authors to referred these faunas to a transitional province [31]. Probably, the invasion of the Boreal stocks resulted from the northward drifting of the blocks in which these faunas dwelt.

Guadalupian (272-262 Ma)
During the Guadalupian Epoch, global eustacy entered a highstand of Permian transgression-regression megacycle, geographic distribution of brachiopods reached an acme subsequent to the Cisuralian. As an outcome of the persistence of the Pangea, the Tethyan faunas became more isolated than in the Kungurian. Similarly, more endemic elements developed in the Boreal and the Gondwanan Realm respectively. In the

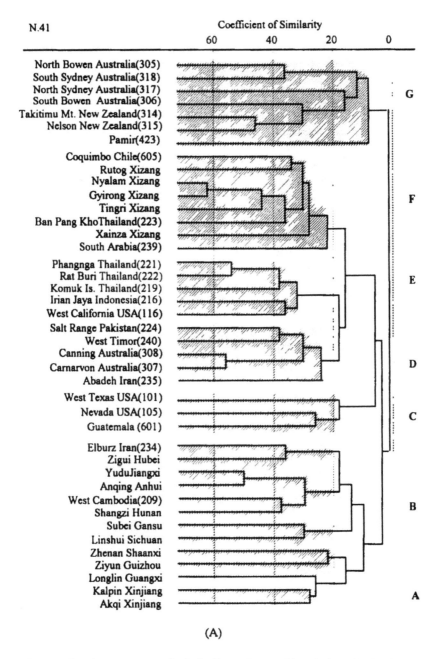

(A)

Figure 4. Dendrograms respectively for Kungurian brachiopods from the Southern Hemisphere (A), and the Northern Hemisphere (B).

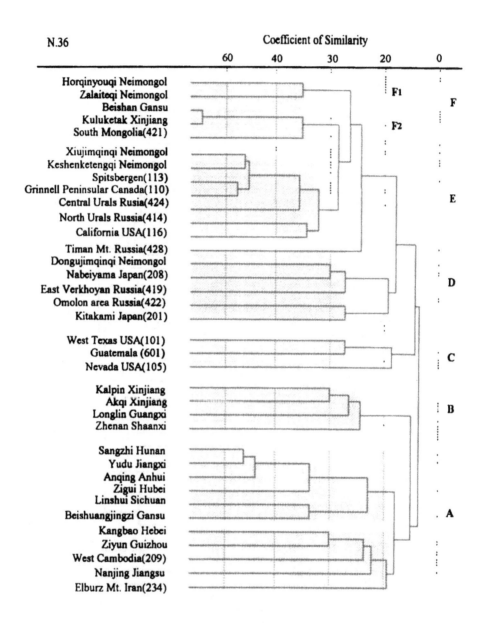

(B)

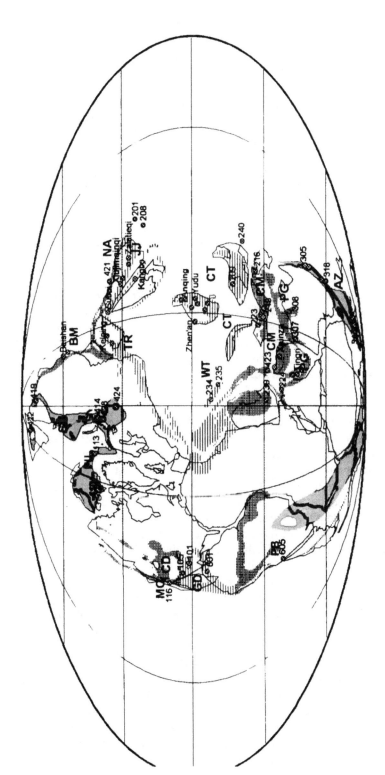

Figure 5. Palaeobiogeographic provincity of Kungurian brachiopods. AZ-Austrazean province, BM-Beishan-Mongol province, CM-Cimmerian province, CD-Cordillera province, CT-Cathaysia province, FU-Franklin-Uralian province, GD-Grandian province, JJ-Jilin-Japan province, MC-McCloud province, NA-Nei Mongol-Amur province, PG-Peri-Gondwana province, PB-Peru-Bolivian province, SB-Siberian provinces, TR-Tarim province, WT-West Tethys province. Legend of shaded areas as in figure 3.

dendrogram for Guadalupian brachiopod faunas from Northern Hemisphere, (Fig. 7A), these clusters are well defined, which represent the Cordillera-Grandian province (cluster B), Zechstein province (cluster C), Arctic and Kazan provinces (cluster E), McCloud province (cluster F), Nei Mongol-Amur province (cluster G) and Cathaysia province (cluster J). The faunas of the cluster H are characterized by a dominance of pro-reef elements and probably, represent a community related to the bioherm and reef in the Tethyan Realm. Among the other clusters of the dendrogram, cluster A is composed of faunas from Hydra Island, Greece and these from Zhen'an of Shaanxi which do not fall clearly in any group since they comprise some unusual pro-reef taxa or are mostly regarded as new forms. The poorly defined cluster D contains the faunas characterized by very low diverse faunas from Omolon of Siberia and its adjacent areas such as Nei Mongol, Jilin and Japan. The grouping of these faunas is not stable as the others.

In the dendrogram of the faunas from Southern Hemisphere (Fig. 7B), the clusters for pro-reef faunas (cluster B), the Cathaysia province (clusters D and E), the Grandian-Cordillera province (cluster F) and the McCloud province (cluster K). The cluster of faunas from South China comprises two branches, one for the faunas from the eastern part of China (cluster D) and the other for those from the Yangtze (cluster E). There are two Gondwanan provinces: the Austrazean province represented by the cluster A and the Peri-Gondwana province, by the cluster L. It is noteworthy that the faunas from Salt Range of Pakistan and Xainza of central Xizang (cluster C) have shifted from the Gondwanan Realm into the Tethyan Realm.

Provinciality of the Guadalupian is essentially the same as that of the Kungurian but there are a few of prominent differences between them. In Northern Hemisphere, two embayments appeared with endemic brachiopod faunas, i.e. the Zechstein Sea and the Kazan Sea. Those from the Zechstein Sea of Late Guadalupian contain such characteristic forms of the Arctic province as *Horridonia*. But the diversity of the Zechstein fauna is much lower than the latter. The brachiopods from Kazan Sea are characterized by a dominance of *Stepanoviella*, an opportunistic form of Permian communities. Widespread pro-reef associations have been reported in low latitude areas. Those from Timor and Sumatra of Indonesia in the east, through Lasala and Chitichum of Xizang, Kiosa of Greece, Sicily of Italy, to Djebt Tebaga of Tunisia in the west form a linear distribution between the Gondwana and the Cimmerian Province. They are mostly found from exotic carbonate blocks, and thus are possibly the relics of carbonate buildup along a series guyots of deep basin [9]. In Southern Hemisphere, the vast embayments of Parana is not hospitable to the brachiopods due to the high salinity.

Lopingian(262-250 Ma)
Marine deposition in the epi-continental seas of Pangea used to be considered as being largely interrupted by a global regression. However, the *Cyclolobus* fauna has been regarded as the Lopingian in age by the ammonoids workers for a long time. This age assignment is confirmed by associated conodont and foraminifer fauns in Salt Range [33]. Accordingly, the Lopingian deposits should exist in the uppermost Permian of peri-Gondwanan and Arctic areas, in which the *Cyclolobus* fauna has been found. The brachiopods are much more common than the ammonoids in the uppermost Permian

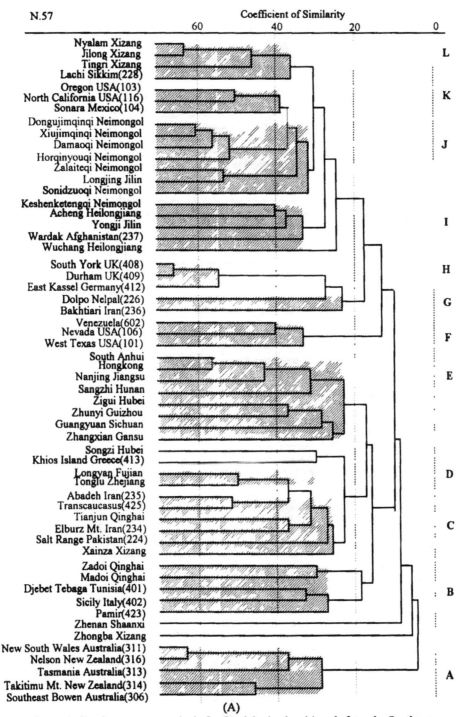

Figure 6. Dendrograms respectively for Guadalupian brachiopods from the Southern Hemisphere (A), and the Northern Hemisphere (B).

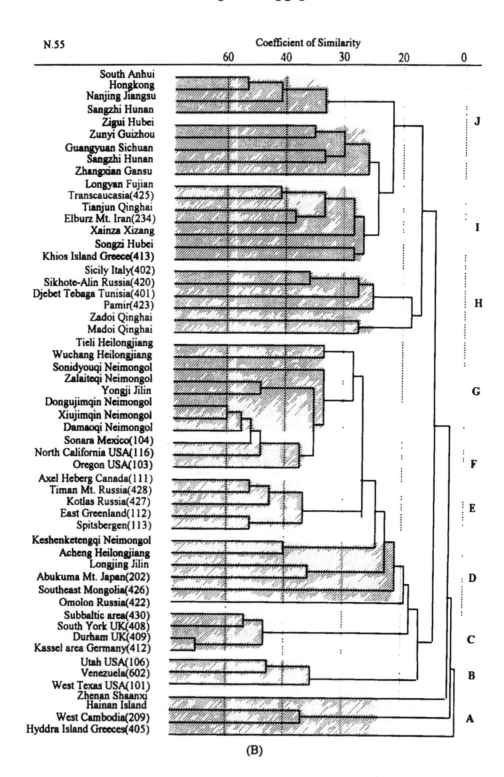

(B)

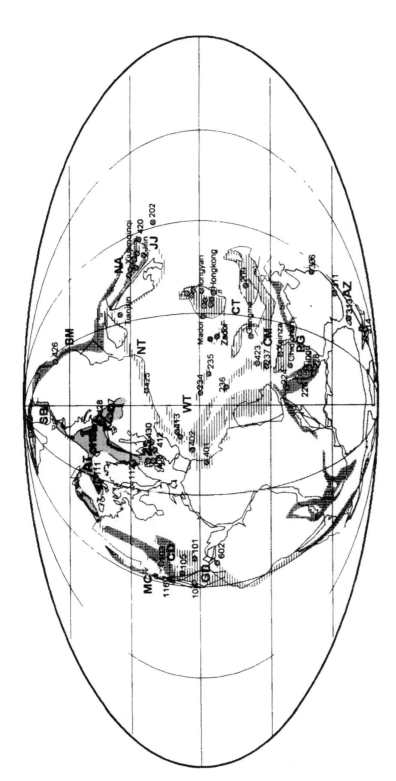

Figure 7. Palaeobiogeographic provincity of Guadalupian brachiopods. AT-Arctic province, AZ-Austrazean province, CM-Cimmerian province, CD-Cordillera province, CT-Cathaysia province, GD-Grandian province, JJ-Jilin-Japan province, KZ-Kazan province, MC-McCloud province, NA-Nei Mongol-Amur province, NT-North Tethys province, PG-Peri-Gondwana province, SB-Siberian provinces, WT-West Tethys province, ZS-Zechstein province. Legend of shaded areas as in figure 3.

formed in the epi-continental seas of Pangea. Nevertheless, the diagnostic characters of the coeval brachiopod fauna of the *Cyclolobus* fauna are still not identifiable except that from Salt Range, where the brachiopods have been dominated by some Tethyan elements. As a consequence of this timing difficulty, very few brachiopod faunas in the areas outside the Tethyan areas can be selected as the Lopingian faunas for cluster analysis.

From the dendrogram of Lopingian faunas (Fig. 9), the clusters of faunas from New Zealand (cluster A) and Peri-Gondwanan province (cluster B) are distinct. Assuming the brachiopods and the *Cyclolobus* fauna from the Martinia Beds of eastern Greenland are coeval, there should be a discrete Arctic province with similar brachiopod taxa to that of the Guadalupian. The Tethyan faunas form a bunch of clusters with progressively changing similarity coefficients. There are the cluster for Cathaysia province (cluster H), the cluster with faunas from Western Tethyan province (Central Iran - Transcaucasus) (cluster E), and the other two ill-defined clusters (clusters C and D). No significant difference in taxonomic composition can be recognized among them though some provinces might have existed during the Lopingian. These provinces can be delineated by a group of endemic forms. For instance, Lopingian brachiopods from southern Alps of western Tethys differ from those elsewhere in having such endemic athyridid genera as *Janiceps*, *Camelicania* and others. Similarly, South China, Transcaucasue and Japan may have been biogeogphically distinguished from one another.

A PANGEA BIOGEOGRAPHIC PATTERN

In summary, the most prominent geographic features constraining the biogeographic pattern of Permian brachiopods varied with epochs. During the Cisuralian, the great-scale glaciation resulted in the Andean, Austrazean, and Perigondwana faunas of the mutually unified Gondwana Realm. The Tethyan faunas covered wider areas and more closely linked to the Boreal faunas than the succeeding epochs. After the Cisuralian, a sharp change took place in the configuration of brachiopod faunas in conformity with the amalgamation of the Pangea. During the Guadalupian, reef and bioherm facies widespread in low latitudinal areas as a consequence of the global highstand eustacy. The disastrous global regression of Lopingian caused further isolation of the endemic faunas. Among them, a Pangea biogeographic pattern throughout post-Artinskian epochs of the Permian is an unique feature.

While the consolidation of Pangea approached at the beginning of the Kungurian, the benthic faunas of Tethys became distinct from other areas. Ross [19] and Leven et al. [11] have emphasized the reorganizing event of fusulinid provinciality at this time. Since the Kungurian, the fusulinids were confined to the areas of low latitude. Two pan-equatorial realms, the American and the Tethyan Realm remained but the Franklian-Uralian Realm disappeared. On the other hand, the brachiopod faunas reflects not only the latitudinal climatic differentiation but also the separation of the pro-Pangea brachiopod faunas from the Tethyan. Three distinct faunal associations are apparent respectively for the brachiopods in the peri-Pangea shelves, the Tethys and those in

N.42

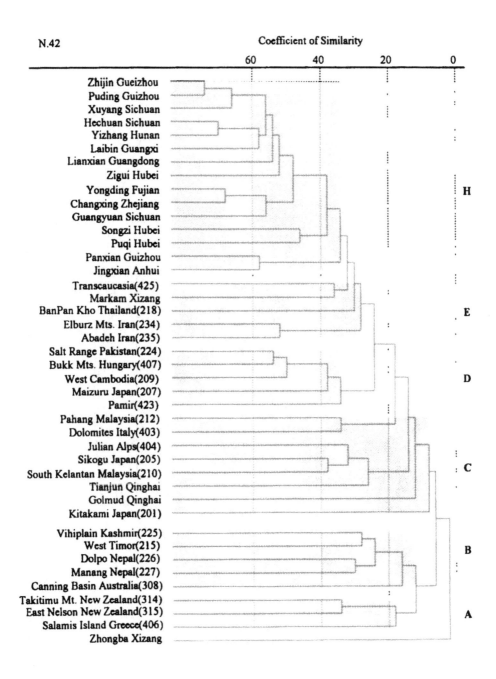

Figure 8. Dendrogram for Lopingian brachiopods

Jin Yugan and Shang Qinghua

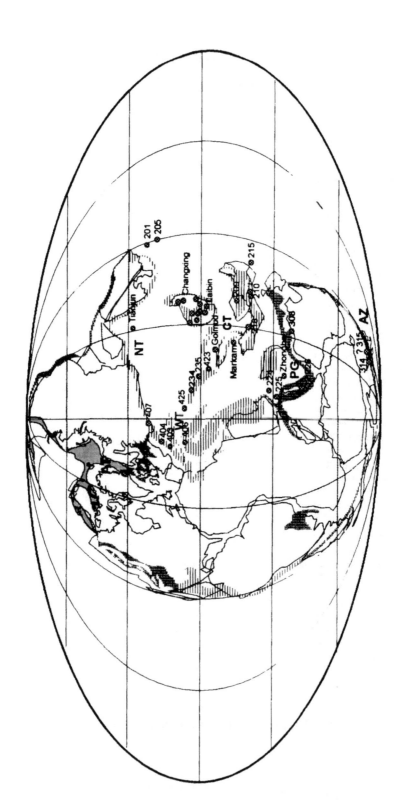

Figure 9. Palaeobiogeographic provincity of Lopingian brachiopods. AT-Arctic province, AZ-Austrazean province, CT-Cathaysia province, NT-North Tethys province, PG-Peri-Gondwana province, WT-West Tethys province. Legend of shaded areas as in figure 3.

between [8]. The brachiopod faunas inhabited in peri-Pangea shelves share an association of pro-Pangea or pro-cratonic taxa such as *Spirifer, Neospirifer, Spiriferella*, the syringothyrids and the buxtonids (Fig. 6). Since they are relic linkages inherited from the cosmopolitan taxa of the Cisuralian, their representatives can be recognized all over the Pangea. Dunbar [5] had pointed out that the dominant species of *Spirifer* from two remote areas, Greenland and Salt Range, are nearly identical. Faunas characterized by a dominance of this association have been reported from the Cordillera, Himalayan, Beishan - Mongol, Peri-Gondwanan, Austrazean and Arctic provinces. It is notable that these faunas extended upward without appreciable evolutionary changes till the end-Permian extinction.

In the meantime, all the pro-Pangea forms retreated from the Tethyan areas extending from southern Alps to southern Japan. Such new endemic groups as the monticuliferids, haydenellids, tyloplectids, cryptothyrids began a remarkably evolutionary diversification from non-dominant elements of Cisuralian faunas in the Tethyan areas. Unlike the conservative brachiopods of Pangean shelves, the Tethyan faunas evolved rapidly, and were profoundly renewed through the initial phase of the end-Permian extinction. They are highly close to one another and hence were suggested as dwellers of restricted-circulated shelf seas.

Between the areas dominated by the pro-Pangea and the Tethyan associations of brachiopods, there are faunas with elements from both associations. The pro-Pangea association is dominant in the low-diversity communities from the clastic beds while the Tethyan association dominates the highly diverse communities from the carbonate beds. These faunas occurred in the Nei Mongol-Amur, Cimmerian, and McCloud provinces as well as in northern Caucasus and northern Japan. These areas had been island arcs fringing the peri-Pangea shelves, but the Tethyan blocks were archipelago-shaped further off the Pangea shelves.

The pro-Pangea association has been regarded as main elements with bipolar distribution. This assumption is in discordance with the common occurrence of this association in the Grandian province. This province is unanimously assigned to the pan-equatorial provinces as tropic forms well developed .

The appearance of pro-Pangea brachiopod association was paralleled by other groups, such as rugose corals [30]. Craton-associated coral faunas nearly extend along entire western margin of Pangea from the Urals to South America *via* Arctic Canada.

Acknowlegements
We are indebted to Prof. A. M. Ziegler and D. Rowley for providing us the data sets of global palaeogeographic atlas. Financial support of National Natural Science Foundation of China (Grant 4967292) and Chinese Academy of Sciences is gratefully acknowledged.

REFERENCES

1. N.E. Archbold and Shi Guangrong. Permian Brachiopod Biogeography of the western pacific in Ralation to Terrane Displacement and Climatic change, *Permophiles* **24**, 10-11 (1994).
2. B.I. Chuvashov, C.B. Foster, G.A. Mizens, J. Roberts and J.C. Claoue-Long. Radiometric (Shrimp)dates for some biostratigraphic horizons and event levels from the Russian and Eastern Australian Upper Carboniferous and Permian, *Permophiles* **28**, 29-36 (1996).
3. G.A. Cooper and R.E. Grant. Brachiopods and Permian Correlations. In: The Permian and Triassic Systems and Their Mutual Boundary. A. Logan and L.V. Hills (Eds). *Canadian Society of Petroleum Geologists* **2**, 572-595 (1973).
4. J.M. Dickins. What is a Pangaea ? . In: *Pangea: Global Environments and Resources. Canadian Society of Petroleum Geologists* **17**, 67-81 (1994).
5. C.O. Dunbar. Permian Brachiopod Faunas of central east Greenland, *Meddelelser om Gronland* **110:3**, 1-169 (1955).
6. R.J. Enkin, Yang Z., Chen Y. and V. Courtillot. Paleomagnetic constraints on the geodynamic history of the major blocks of China from the Permian to the Present, *Am. Mineral.* **97**, 13953-13989 (1992).
7. N. Fantini Sestini. The Geology of the Upper Djadjerud and Lar Valleys (North Iran). II Palaeontology. Brachiopods from Dorud Formation. *Rivista Italiana di Paleontologia e Stratigrafia (Milan)* **71:3**, 773-789 (1965).
8. Jin Yugan. On the Paleoecological Relation betweeen Gondwana and Tethys Fauns in the Permian of Xizang. In: *Geological and Ecological Studies of Qinghai-Xizang Plateau*, **1**, pp. 171-178. Beijing Science Press (1981).
9. Jin Yugan. Permian brachiopods and Palaeogeography of the Qinghai-Xizang(Tibet) Plateau. *Palaeontologia Cathayana*. **1(2)**:19-56 (1985).
10. Jin Yugan. Bruce R. Wardlaw, B.F. Glenister and G.V. Kotlyar. Permian chronostratigraphic subdivisions. *Episodes* **20(1)**: (1997).
11. E.Ya. Leven, M.F. Bogoslovskaya, V.G. Gnelin, T.A. Grunt, T.B. Leonova, and A.N. Reimers. Reorganization of marine biota during the mid-Early Permian epoch, *Stratigraphy and Geological Correlation*, **4:1**, 57-66 (1996).
12. Li Li and Gu Feng. Permian brachiopod. In: *Palaeontological Atlas of North Chian. I. Palaeozoic Volume: Inner Mongolia*, pp. 228-305. Geol. Publ. House, Beijing (1976).
13. Li Li. and Li Wenguo. Carboniferous and Permian Brachiopods. In: *Palaeontological Atlas of Northeast China. I. Palaeozoic Volume*, pp. 327-428. Geol. Publ. House, Beijing (1980).
14. Liang, Wenping.. The Lengwu Formation of Permian and its Brachiopod Fauna in Zhejiang Provice, *Geological Memoirs*, Ser.2, **10**, 1-522 (1990).
15. Nie Shangyou, D.B. Rowley and A.M. Ziegler. Constraints on the locations of Asian micro-continents in the Palaeo-Tethys during the Late Palaeozoic, *Geol. Soc. Mem.* **12**, 397-409 (1990).
16. L.I. Popeko, B.A. Nataliin, G.V. Belyaeva, G.V. Kotlyar and G.R. Shishkina. Paleozoic Biogeographic Zones and Geodynamics of The Southern far east of Russia, *Geol. Of Pac. Ocean* **10:5**, 817-830 (1994).
17. J. Roberts, J.C. Claoue'-Long and C.B. Foster. SHRIMP zircon dating of the Permian System of eastern Australia, *Australian J. Earth Sci.* **43**, 401-421 (1996).
18. C.A. Ross and J.R.P. Ross. Late Paleozoic sea levels and depositional sequences. In: *Timing and depositional history of eustatic sequences, Constraints on seismic stratigraphy*. C.A. Ross and D. Haman (Eds). *Cushman Foundation for Foraminiferal Research, Special Publication* **24**, 137-149 (1987).
19. C.A. Ross. Evolution of Fusulinacea (Protozoa) in Paleozoic space and time. In: *Historical biogeography, plate tectonics, and the changing environment*. Gray and Boucot (Eds). pp.215-227 (1979).

20. C.A. Ross. Permian Fusulinaceans . In: *Permian of the Northern Pangea*, 1. P.A. Scholle, T.M. Peryt & D.S. Ulmer-Scholle (Eds), pp. 167-186. Springer-Verlag, Berlin (1995).
21. C.R. Scotese and R.P. Langford. Pangea and the Paleogeography of the Permian. In: *Permian of the Northern Pangea*, 1. P.A. Scholle, T.M. Peryt and D.S. Ulmer-Scholle (Eds), pp. 3-20. Springer-Verlag, Berlin (1995).
22. A.M.C. Sengor, B.A. Natalin and V.S. Burtman. Evolution of the Altaid tectonic collage and Palaeozoic crustal growth in Eurasia, *Nature* 364, 299-307 (1993).
23. Sheng Jinzhang and Jin Yugan. Correlation of Permian deposits in China. In:*Permian stratigraphy, Environments and Resources Vol. 1:Palaeontology & Stratigraphy.* Jin, Utting and Wardlaw (Eds). *Palaeoworld* 4, 14-113 (1994).
24. Shi Guangrong Multivariate data analysis in Palaeoecology and Palaeobiogeography-arevien, *Palaeogeography, Palaeoclimatology, Palaeoecology* 105, 199-234 (1993).
25. Shi Guangrong. and N.W. Archbold. A quantitative analysis on the distribution of Baigendzhinian-Early Kungurian (Early Permian) brachiopod faunas in the western Pacific region, *Journal of Southeast Asian Eadrth Sciences* 11:3, 189-205 (1995).
26. F.G. Stehli and R.E. Grant. Permian Brachiopods from Axel Heiberb Island, Canada, and an Index of Sampling Efficiency, *Journal of Paleontology* 45:3, 502-521 (1971).
27. F.G. Stehli. Possible Permian Climatic zonation and its Implications, *American Journal of Science* 255, 607-618 (1957).
28. F.G. Stehli. Taxonomic Diversity Gradients in Pole Location: The Recent Model, In: *Evolution and Environment.* pp. 163-227. New Haven: Yale Univ. Press. (1968).
29. F.G. Stehli. Tethyan and Boreal Permian Faunas and their significance, *Smithsonian Contributions to Paleobiology* 3, 337-345 (1971).
30. C.H. Stevens. The Early Permian Thysanophyllum coral belt: Another clue to Permian plate tectonic reconstructions, *Geological Sociaty America Bulletin* 93, 798-803 (1982).
31. Tazawa J.. Middle Permian Brachiopod Biogeography of Japan and Adjacent Regions in East Asia. In: *Pre-Jurassic Geology of Inner Mongolia, China.* pp. 213-230 (1991).
32. V.I. Ustritskiy. Principal Stages in the Permian Evolution of Asian Marine Basins and Brachiopod Fauna, *International Geology Review* 4:4, 415-426 (1962).
33. B.R. Wardlaw and K.R. Pogue. The Permian of Pakistan. In: *Permian of the Northern Pangea*, 2. P.A. Scholle, T.M. Peryt & D.S. Ulmer-Scholle (Eds), pp. 215-225. Springer-Verlag, Berlin (1995).
34. J.B. Waterhouse and G.F. Bonham-Carter. Global Distribution and Character of Permian Biomes Based on Brachiopod Assemblages, *Canadian Journal of Earth Sciences* 12:7, 1085-1146 (1975).
35. T.E. Yancey. Permian Marine Biotic Provinces in North America, *Journal of Paleontology* 49:4, 758-766 (1975).
36. Zhang Shouxing and Jin Yugan. Upper Paleozoic brachiopoda from the Mount Jolmolungma region. In: *A report of scientific expedition in the Mount Jolmolungma region (1966-1968) (Paleontology, fasc. 2).* pp.159-242. Sci. Press, Beijing (1976).
37. A.M. Ziegler Phytogeographic patterns and continental configurations during the Permian Period, *Geol. Soc. Mem.* 12, 363-379 (1990).
38. A.M. Ziegler. M.L. Hulver and D.B. Rowley. Permian world topography and climate. In: *Late Glacial and Post-Glacial Environmental Changes-Quaternary, Carboniferous-Permian and Proterozoic.* I.P. Martini (Ed).Oxford University Press, New York (1996).

Proc. 30ᵗʰ Int'l. Geol.. Congr., Vol. 12, pp. 54-66
Jin and Dineley (Eds)
© VSP 1997

Notes on the Historical Biogeography of the Osteoglossomorpha (Teleostei)

LI GUOQING

Institute of Vertebrate Paleontology and Paleoanthropology, Chinese Academy of Sciences, P. O. Box 643, Beijing 100044, China / Department of Biological Sciences and Laboratory for Vertebrate Paleontology, University of Alberta, Edmonton, Alberta, Canada T6G 2E9

Abstract

The superorder Osteoglossomorpha is a monophyletic group with a relatively well-preserved fossil record extending from Late Jurassic to Oligocene in the freshwater deposits of the world. Both cladistic vicariance and dispersal events, combined with extinctions, could have contributed to establishing of the Recent transoceanic distribution of this teleostean group. Based on the age and distribution of the group, the most likely ancient landmass available as an ancestral distribution area of Osteoglossomorpha is Pangea; that is, the early development of Osteoglossomorpha in Pangea was probably completed by the Middle to the Late Jurassic, and the earliest of the main lineages of this group had already enlarged their distribution to most parts of Pangea before its final fragmentation. It is also suggested that the ÝLycopteriformes is a clade endemic only to East Asia as East Asia was isolated from Euramerica and all the southern continents during Late Jurassic to Early Cretaceous time, and the order has never been found elsewhere. Fossil evidence from the Lower Tertiary of all the main continents except Antarctica suggests an "Early Tertiary global diversification" of Osteoglossomorpha. The Recent distribution of this freshwater fish group has probably resulted from extinction of the Osteoglossiformes from North America, East Asia, and Europe, and the Hiodontiformes from East Asia and far western North America (now the Rocky Mountains area).

Keywords: Osteoglossomorpha, Pangea, trans-oceanic distribution, vicariance, extinction

INTRODUCTION

The superorder Osteoglossomorpha, defined by Greenwood *et al.* in 1966 [15], has been involved in numerous systematic studies in the years since [11, 13-14, 18-19, 22-24, 27-28, 36, 47]. However, the Recent transoceanic distribution of this teleostean fish group (Fig. 1) has not yet been well resolved. Nelson [28] believed that Africa could be the center of the ancestral distribution of osteoglossomorphs; accordingly, the presence of Hiodontidae in North America could be secondary and of relatively late occurrence. Greenwood [13] and Zhang and Zhou [54], upon the establishment of the relationship between Lycoptera and Hiodontidae, suggested that osteoglossomorphs might have been distributed ancestrally in East Asia, later reaching North America and the southern continents. Patterson [34-35] raised the possibility of marine dispersal of the ancestral lineage of Osteoglossoidei based on the discovery of *Brychaetus* in marine deposits. I

[18] further discussed the historical biogeography of Hiodontidae based on recognition of a later Early Cretaceous hiodontid *Yanbiania*, believing that Hiodontidae were originally distributed in north-eastern Asia. All these hypotheses considered dispersal to be the major factor affecting osteoglossomorph biogeography.

More recently, the vicariance approach [4, 8, 11, 30-31, 38, 40, 52] has been applied to the transoceanic relationships of Osteoglossomorpha. In an extreme example, Gayet [6] believed that the present distribution of osteoglossomorphs could be the result of the fragmentation of a hypothetical "lost Pacifica" [32].

Attempts to explain the trans-Pacific relationships of North American fossil and Recent osteoglossomorphs have also been made by Grande [8, 10], who proposed a "general biological pattern" involving the early Tertiary of western North America; and by Wilson and Williams [53], who suggested that the present endemic distribution of hiodontids may be a remnant of a former widespread northern-hemispheric distribution.

It is now evident that the historical biogeography of Osteoglossomorpha is much more complicated than has been discussed or expected. Biogeographers and ichthyologists could be misled if they rely only on one model while neglecting the others; as well, they may be misled, even if they have considered both dispersal and vicariance of extant species, but ignore the fossil evidence.

Philosophical Difference between Dispersal and Vicariance Models
Grande [8], Bănărescu [1], Wilson and Williams [53, partly], for example, recently summarised the argumentation between dispersal and vicariance models. The main distinctions between these two models are: 1) dispersal looks upon the sequence of fossil records and uses fossils as evidence in reconstructing the historical biogeography of a taxon, while vicariance emphasises the incompleteness of fossil records, and often rejects fossil evidence; 2) in explanations relying on dispersal, it is assumed that enlargement and shifting of the distribution area of a taxon is common and requires the presence of an ancestral distribution area (center of origin), while explanations relying on vicariance assume that fragmentation of whole biotas is common, and centers of origin are ignored. A critical survey of the principles of both dispersal and vicariance models affirms that both models are logical. Neither dispersal nor vicariance is likely without movement, but the dispersal model emphasises the movement of a lineage from one area to another (actively or passively), while the vicariance model stresses the subdivision and possible movement (drift) of the land mass carrying that lineage. In practice, both processes might have functioned alternatively during the biogeographical history of a lineage. This appears to be true with the historical biogeography of the Osteoglossomorpha, that is, both vicariance and dispersal events, combined with extinctions, could have contributed to the establishment of the Recent distribution of Osteoglossomorpha. The known fossil record of this group, from the Upper Jurassic to the Oligocene and from all the continents except Antarctica, provides studies of this group with the key advantage of "time-control" [8].

Contributions of Fossil Evidence toward Paleobiogeographic Reconstruction

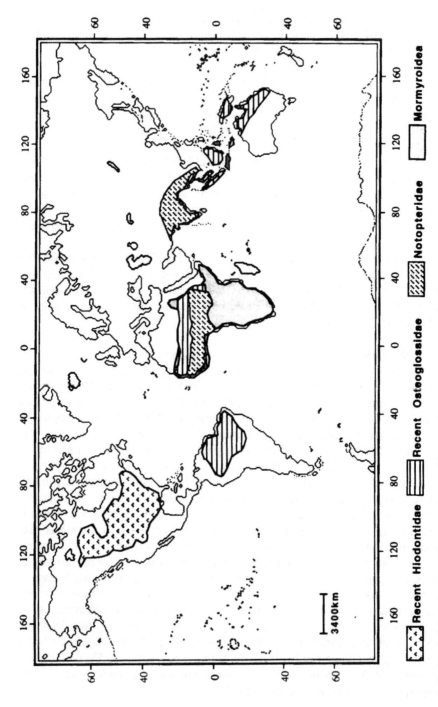

Figure 1. Map showing the modern geographic distribution of Osteoglossomorpha.

Grande [8, 11] and Wilson and Williams [53] recently discussed the importance of fossil evidence in the reconstruction of paleobiogeography. In agreement with Grande [8], I consider four basic contributions of fossil evidence toward the solution of paleobiogeographic problems in this paper: (1) fossils provide morphological data on taxa in addition to those provided by Recent species, (2) fossils provide additional taxa that increase the known biogeographic range of a lineage, (3) fossils establish a minimum age for a taxon, and (4) fossil demonstrate area relationships that are not recognisable from later fossils or from extant biotas. All these functions are applicable in reconstructing the osteoglossomorph historical biogeography except (4), which needs to be confirmed by a "repeating pattern" [11] of another lineage in the biota.

THREE HYPOTHESES ON OSTEOGLOSSOMORPH PALEOBIOGEOGRAPHY

During the last two decades, a number of extinct taxa has been added to the fossil record of the Osteoglossomorpha [6, 9, 17-19, 21-23, 42, 47], all of them suggesting transoceanic relationships of the superorder. Recent study of both the extant and fossil osteoglossomorphs [24] further improves understanding of the evolutionary history of the entire lineage as shown in Fig. 2, which briefly summarises the stratigraphic and phylogenetic history (vertical distribution) of the Osteoglossomorpha. Plotting taxa known from different periods in Fig. 2 on maps results in Figs. 1, 3, and 4, which show the change of distribution of Osteoglossomorpha from the Early Cretaceous to the Recent. A comprehensive interpretation of these maps suggests the following three hypotheses.

1. Early Evolution of Osteoglossomorpha Occurred in Pangea
As shown in Fig. 2, the known earliest record of Osteoglossomorpha is the Late Jurassic or Early Cretaceous *Lycoptera*. In addition, fossil teleosts considered to represent the main lineages of Osteoglossomorpha have been reported from the Lower Cretaceous of the north-eastern part (*Yanbiania*) [18] and the central and southern parts of East Asia (e.g., *Paralycoptera*) [5], from the Albian to Cenomanian (*Chandlerichthys*) [9] and Campanian to Maastrichtian (*Cretophareodus and Ostariostoma*) [12, 21, 43] of North America, and from the Aptian of South America (*Laeliichthys*) [42]. According to Taverne [49] and Bonde [3], a Cenomanian genus of Osteoglossomorpha (*Kipalaichthys*) is known from Africa. A possible osteoglossomorph (*Koonwarria*) [50, pers. obv., further study required] has also been found in the Lower Cretaceous of southeastern Australia. Judging from the preserved skeletal characters, these genera seem unlikely to represent the earliest lineages of Osteoglossomorpha because they already share most of the osteoglossomorph synapomorphies with their extant relatives [19, 24]. Some of them (e.g., *Yanbiania*) are distinguished from their extant relatives only by relatively minor features.

As regards the distributions of the above mentioned genera, they occupied most of the main land masses of the Laurasia and Gondwana that were breaking up. It is very difficult to interpret these data by suggesting the early development of Osteoglossomorpha in "Africa" [28] or in "East Asia" [5, 13, 54]. These two hypotheses

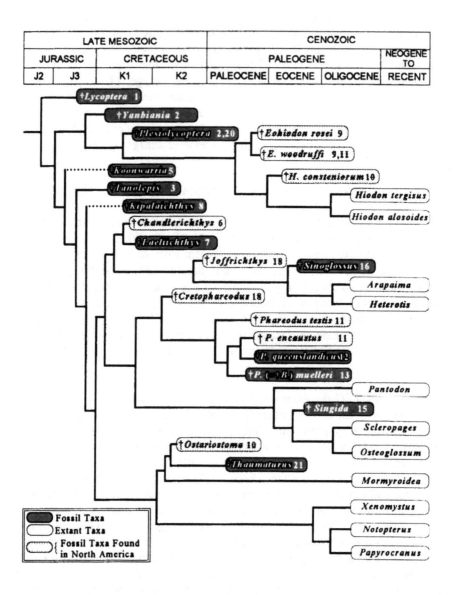

Figure 2. Diagram showing the hypothetical evolutionary history of Osteoglossomorpha. Numbers following all taxa represent localities which are plotted in Figs. 3 and 4. Note that all the main lineages of this fish group had a North American distribution. *E., (Eohiodon); H., (Hiodon); P., (Phareodus); B., (Brychaetus).*

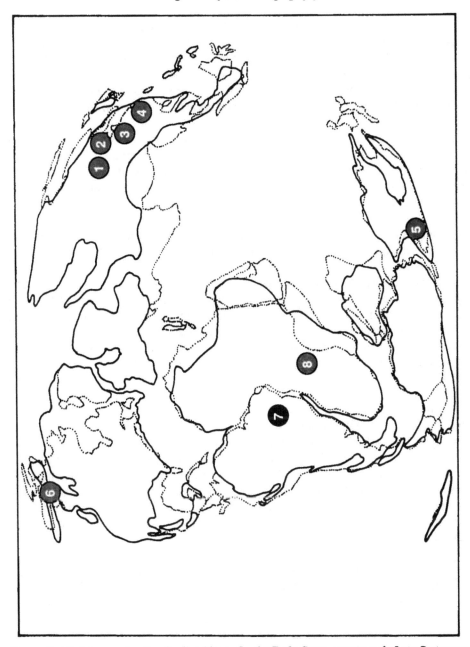

Figure 3. Sketch map showing fossil evidence for the Early Cretaceous to early Late Cretaceous distribution of Osteoglossomorpha. 1. *Lycoptera*, 2. *Yanbiania*, 3. *Tanolepis*, 4. *Paralycoptera*, 5. *Koonwarria*, 6. *Chandlerichthys*, 7. *Laeliichthys*, and 8. *Kipalaichthys*.

demand that the ancestors of Osteoglossomorpha would have to make at least four transoceanic dispersals from Africa to South America, North America, Australia, and East Asia; or, from East Asia to North America, Australia, South America, and Africa during the time between the Late Jurassic and the Late Cretaceous. The hypothesis of a "lost Pacifica" [6] is even more vulnerable because it needs at least three vicariance events and drifting in three directions followed by collisions of the fragments of the so-called "lost Pacifica" with East Asia, North America, and South America, and at least one transoceanic dispersal from South America to Africa during the same time. Besides, there is no sufficient geological evidence to document the existence of such a continent.

Thus, the ancestors of Osteoglossomorpha were most probably much more plesiomorphic than *Lycoptera, Yanbiania, Paralycoptera, Chandlerichthys, Koonwarria, Laeliichthys,* and *Kipalaichthys*. They could well have existed much earlier and possibly had a much more widespread distribution than any of these Late Jurassic or Early Cretaceous genera.

The most likely ancient landmass available as an ancestral distribution area of Osteoglossomorpha is Pangea [51], a supercontinent thought to be present from the end of the Permian to the end of the Middle Jurassic. It seems probable that the early development of Osteoglossomorpha had taken place in Pangea by the Middle to the Late Jurassic time, and that the earliest of the main lineages of this group had already enlarged their distribution to most parts of Pangea before its final break-up (Fig. 3). The subsequent fragmentation of Pangea established the framework of a global transoceanic distribution pattern of the Osteoglossomorpha. Fossil evidence documenting this hypothesis comes from India [16] and all the main continents except Antarctica (Figs. 3 and 4).

2. *Lycopteriformes are Endemic to East Asia*
Recent study of fossil osteoglossomorphs also provides important information for identifying the endemism of the Lycopteriformes, a stem group of Osteoglossomorpha [19, 24]. Based on the available fossil record of *Lycoptera* [25], the ancestral lineage of lycopteriformes probably entered East Asia before the Late Jurassic, inhabiting the freshwater of North China, Mongolia, and eastern Siberia (east of Baikal Lake), an area ranging from 35°-60°N and 95°-135°E [18]. This distribution of Lycopteriformes, including *Lycoptera* and possibly *Tongxinichthys* [25], lasted from at least the Late Jurassic to the Early Cretaceous.

As far as is known to me, there is no certain record of Lycopteriformes outside of East Asia. It seems to me that this distribution pattern of lycopteriformes can not be simply asserted as owing to the incompleteness of the fossil record. It is also hard to imagine the extinction or the almost global disappearance of the Lycopteriformes, which demands at least seven extinction events of lycopteriforms from North America, South America, Europe, India, Africa, Australia, and Antarctica. If so, it is unlikely that these extinction events left no fossil record of the Lycopteriformes in all those continents.

Accordingly, I suggest another possibility: that Lycopteriformes are endemic only to East Asia. This hypothesis is basically compatible with the Late Mesozoic plate tectonic framework. According to Smith and Briden [44], Olsen [33], Smith and Hurley et al. [45], Briggs [2], and Smith and Smith et al. [46], during Late Jurassic to Early

Cretaceous time, East Asia was separated from Euramerica by the epicontinental Turgai Sea and Turgai Strait, which existed until the late Eocene, and from all the southern continents by the eastern part of the Tethys Sea. In addition, the connection between western North America and East Asia was not completely established at that time. These divisions might have functioned as barriers that prevented freshwater Lycopteriformes from expanding over North America, Europe, and all the southern continents. This idea seems to be easily refuted if fossils of Lycopteriformes are ever found outside East Asia. However, I predict that they never will be based on the principle of predictability suggested by Grande [8, 11].

3. The Recent Relict Distribution of Osteoglossomorpha Resulted from Extinction

As already mentioned, the early development of Osteoglossomorpha was probably completed in Pangea by the Middle to the Late Jurassic, before its final disintegration. It is thus evident that the fragmentation of the supercontinent Pangea established a global distribution pattern of the Osteoglossomorpha. Further investigation suggests that this freshwater teleostean group came up to a "period of global diversification" in the Early Tertiary, in which osteoglossomorphs were thriving on almost all the main continents. This suggestion is supported by a number of well-studied fossil osteoglossomorphs from the Lower Tertiary of various localities in the world (Fig. 4). Available evidence includes seven species from western North America (*Eohiodon rosei, Eohiodon woodruffi, Hiodon consteniorum, Joffrichthys symmetropterus, Phareodus encaustus, Phareodus testis*, and *Ostariostoma wilseyi*), three species from East Asia (China) {*Sinoglossus lushanensis*, [46]; *Plesiolycoptera (Gobihiodon) parvus*, [48], and *?Phareodus* sp.}, two species from western *Europe [Phareodus (=Brychaetus) muelleri* and *Thaumaturus*], and one species each from eastern Australia (*Phareodus queenslandicus*), southeastern Asia [*Musperia radiata* (Heer) Sanders, 41], eastern Africa (*Singida*), and South America (*Phareodusichthys tavernei*) [7]. Fossil scales of Osteoglossiformes have also been reported from the Lower Tertiary of India [16].

When Fig. 1 is compared with Fig. 4, I find that the Recent distribution of Osteoglossomorpha is much smaller than it was in the Early Tertiary. On the map showing the Recent transoceanic distribution of the Osteoglossomorpha, Osteoglossiformes have disappeared from East Asia, North America, and Europe; Hiodontiformes that had been present in East Asia and far western North America (an area covering the Rocky Mountains) have disappeared from the freshwater system of East Asia and far western North America, and are endemic only to the freshwaters of the inland North America east to the Rocky Mountains. This phenomenon indicates that extinction must have been involved in the Cenozoic evolution of the Osteoglossomorpha.

According to the fossil record, the "period of global diversification" of the Osteoglossomorpha in the Early Tertiary might have lasted till the end of the Oligocene. It was then interrupted by a series of Neogene plate interactions and dramatic plate motions that have resulted in the vertical uplift of western North America (Rocky Mountains), the formation of the Alps (which resulted from collisions between northern Africa and southern Europe) and Himalayas (which resulted from collisions between

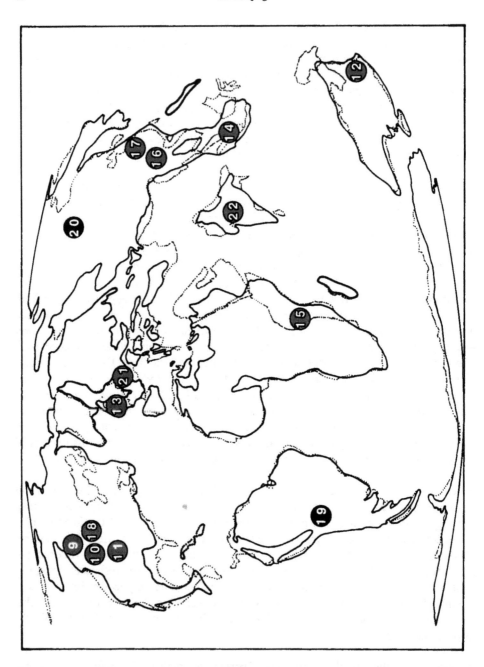

Figure 4. Sketch map showing the Early Tertiary distribution of Osteoglossomorpha. 9 *Eohiodon*, 10 *Ostariostoma, Hiodon consteniorum*, 11 *Phareodus, Eohiodon*; 12 *Phareodus queenslandicus*, 13 *Phareodus (=Brychaetus) muelleri*, 14 *Musperia*, 15 *Singida*, 16 *Sinoglossus*, 17 ? *Phareodus* sp., 18 *Joffrichthys*, 19 *Phareodusichthys*, 20 *Plesiolycoptera (Gobihiodon) parvus*, 21 *Thaumaturus*, 22 fossil scales of Osteoglossiformes.

India and Eurasia). These Cenozoic (Neogene) large-scale tectonic movements greatly deformed the pre-Cenozoic earth's surface, which remoulded particularly the general configuration of the northern continents, and then changed the pre-Neogene freshwater systems of Eurasia and far western North America (now the Rocky Mountains area), the climatic zones, and the habitat environment to which the osteglossomorph fishes had been adapted. All these changes functioned together, resulting in the extinction of the Osteoglossomorpha from East Asia, Europe, and far western North America.

Some Difficult Problems
The present paleobiogeographical interpretation of fossil osteoglossomorphs is limited by some difficult problems, such as the influence of migration. Among the extant fishes, migration refers to those diadromous species that live regularly part of their lives in fresh waters and part in oceans. Two patterns, anadromous and catadromous, are recognised. These two diadromous patterns must have also functioned in extinct fishes, resulting in a mixed fossil fish assemblage (or fauna) consisting of both freshwater and marine species. The passive migration of both dead and living fishes through water transport and biological activities can also result in an assemblage that consists of different species living at varying distances or far from their ultimate site of deposition and burial. This makes it difficult to distinguish between in situ species and drifted taxa in the reconstruction of fossil fish assemblage and paleobiogeography. Moreover, further study is needed to answer questions such as why Hiodontiformes seem to be restricted to the northern hemisphere but Osteoglossiformes have a globally widespread distribution, and the role, if any, played by Antarctica in the evolutionary history of the Osteoglossomorpha.

CONCLUSIONS

The hypothetical ancestral osteoglossomorphs were probably much more plesiomorphic than any of the known genera, such as *Lycoptera, Yanbiania, Paralycoptera*, and *Laeliichthys*; they could well have existed much earlier and probably had a much more widespread distribution than any of these genera. Consequently, it is likely that the early diversification of the Osteoglossomorpha had occurred in Pangea by the Middle to the Late Jurassic time, before the final break-up of that supercontinent. Lycopteriformes, however, are probably endemic only to East Asia. The fossil record documents an "Early Tertiary global diversification of the Osteoglossomorpha." The later Cenozoic large-scale tectonic movements might have resulted in the extinction of this teleostean group from East Asia, Europe, and far western North America.

Acknowledgements

This paper is derived from the third part of the last chapter of my PhD. Thesis [19]. My best thanks are due to Mark V. H. Wilson, University of Alberta, for financial support, discussion, and valuable suggestions on this research; to Richard C. Fox, and Joseph S. Nelson, University of Alberta, for discussion, constructive suggestions, and helpful review of the early version of this paper. I also thank Lance Grande, Field Museum of Natural History, and David Bardack, University of Illinois, for valuable comments and helpful discussion. This research was supported mainly by Natural Sciences and Engineering Research Council operating grant A9180 to Mark V. H. Wilson, and the scholarship to Li Guo-Qing from the University of Alberta.

REFERENCES

1. P. Bănărescu. Zoogeography of fresh waters. 1: General Distribution and Dispersal of Freshwater Animals. Aula-Verlag Wiesbaden, Germany, pp.511 (1990).
2. J. C. Briggs. Introduction to the zoogeography of North American fishes. In: *The Zoogeography of North American Freshwater Fishes*. C. H. Hocutt and E. O. Wiley (Eds.). pp. 1-16. John Wiley and Sons. Inc., New York (1986).
3. N. Bonde. Osteoglossids (Teleostei: Osteoglossomorpha) of the Mesozoic. Comments on their interrelationships. In: *Mesozoic Fishes: Systematics and Paleoecology*. G. Arratia and G. Viohl (Eds.). pp. 273-284. München (Dr. Pfeil), Germany (1995).
4. D. R. Brooks. Parsimony analysis in historical biogeography and coevolution: methodological and theoretical update. *Syst. Zool.* **39**, 14-30 (1990).
5. Chang (=Zhang) Mi-mann and Chou (=Zhou) Chia-chien. On Late Mesozoic fossil fishes from Zhejiang Province, China. *Inst. Vert. Palaeont. Palaeoanthr., Academia Sinica, Memoir*. **12**, 1-59 (1977).
6. M. Gayet. Consideraciones preliminares sobre la paleobiogeografia de les Osteoglossomorpha. *IV Con. Latinoamer. Paleont., Bolivia*. **1**, 379-398 (1987).
7. M. Gayet. "Holostean" and teleostean fishes of Bolivia. *Revista Tecnica de Ypeb*. **12**, 453-494 (1991).
8. L. Grande. The use of paleontology in systematics and biogeography, and a time control refinement for historical biogeography. *Paleobiology* 1985, **11**, 234-243 (1985).
9. L. Grande. The first articulated freshwater teleost fish from the Cretaceous of north America. *Palaeontology* **29**, 365-371 (1986).
10. L. Grande. The Eocene Green River Lake system, Fossil Lake, and the history of the North American fish fauna. In: *Mesozoic / Cenozoic vertebrate paleontology: Classic localities, contemporary approaches. 28th IGC Fieldtrip Guidebook T322 (American Geophysical Union)*. J. Flynn (Ed.). pp. 18-28 (1989).
11. L. Grande. Repeating patterns in nature, predictability, and "impact" in science. In: *Interpreting the Hierarchy of Nature*. L. Grande and O. Rieppel (Eds.). pp. 61-84. Academic Press, San Diego (1994).
12. L. Grande and T. M. Cavender. Description and phylogenetic reassessment of the monotypic Ostariostomidae (Teleostei). *J. Vert. Paleont*. **11**, 405-416 (1991).
13. P. H. Greenwood. On the genus Lycoptera and its relationship with the family Hiodontidae (Pisces, Osteoglossomorpha). *Bull. Brit. Mus. (Nat. Hist.), Zool*. **19**, 257-285 (1970).
14. P. H. Greenwood. Interrelationships of osteoglossomorphs. In: *Interrelationships of Fishes*. P. H. Greenwood, R. S. Miles, and C. Patterson (Eds.). pp. 307-332. Academic Press, London (1973).
15. P. H. Greenwood, D. E. Rosen, S. H. Weitzman and G. S. Ourers. Phyletic studies of teleostean fishes, with a provisional classification of living forms. *Bull. Amer. Mus. Nat. Hist.* **131**, 339-456 (1966).
16. S. L. Hora. On some fossil fish-scales from the Inter-Trappean Beds at Deothan and Kheri, central provinces. *Rec. Geol. Sur. India*. **73**, 267-294 (1938).

17. Jin Fan. A new genus and species of Hiodontidae from Xintai, Shandong. *Vert. PalAsia.* 29, 46-54 (1991) [=*Tanolepis* Jin, 1994, *Vert. PalAsia.* 32, 70 (1994)].

18. Li Guo-qing. A new genus of Hiodontidae from Luozigou Basin, east Jilin. *Vert. PalAsia.* 25, 91-107 (1987).

19. Li Guo-qing New Osteoglossomorphs (Teleostei) from the Upper Cretaceous and Lower Tertiary of North America and Their Phylogenetic Significance. Ph. D. Thesis, University of Alberta Canada, 290pp. (1994).

20. Li Guo-qing. Systematic position of the Australian fossil osteoglossid fish ÝPhareodus (=Phareoides) queenslandicus Hills. *Mem. Queensl. Mus.* 37, 287-300 (1994).

21. Li Guo-qing. A new species of Late Cretaceous osteoglossid (Teleostei) from the Oldman Formation of Alberta, Canada, and its phylogenetic relationships. In: *Mesozoic Fishes: Systematics and Paleoecology.* G. Arratia and G. Viohl (Eds.). pp. 285-298. München (Dr. Pfeil), Germany (1995).

22. Li Guo-qing and M. V. H. Wilson. An Eocene species of Hiodon from Montana, its phylogenetic relationships, and the evolution of the postcranial skeleton in the Hiodontidae (Teleostei). *J. Vert. Paleont.* 14, 153-167 (1994).

23. Li Guo-qing. and M. V. H. Wilson. The discovery of Heterotidinae (Teleostei: Osteoglossidae) from the Paleocene Paskapoo Formation of Alberta, Canada. *J. Vert. Paleont.* 16, 198-209 (1996).

24. Li Guo-qing and M. V. H. Wilson. Phylogeny of Osteoglossomorpha. In: *Interrelationships of Fishes.* M. Stiassny, L. Parenti, and G. David Johnson (Eds.). pp. 163-174. Academic Press, Inc., Orlando, Florida (1996).

25. Ma Feng-zhen. A new genus of Lycopteridae from Ningxia, China. *Vert. PalAsia.* 18, 286-295 (1980).

26. J. Müller. Fossile Fische. In: *Reise in den aussersten norder und osten Sibiriens wahrend der Jahre 1843 und 1844 (Erster Band).* A. Th. Von. Middendorff (Ed.). pp. 260-264. Buchdruckerei der Kaiserlichen Akademie der Wissenschaften, St. Petersburg (1847).

27. G. J. Nelson. Gill arches of teleostean fishes of the division Osteoglossomorpha. *J. Linn. Soc. London (Zool.)* 47, 261-177 (1968).

28. G. J. Nelson. Infraorbital bones and their bearing on the phylogeny and geography of osteoglossomorph fishes. *Amer. Mus. Novit.* 2394, 1-37 (1969).

29. G. J. Nelson. Observations on the gut of the Osteoglossomorpha. *Copeia* 1972, 325-329 (1972).

30. G. J. Nelson and N. I. Platnick. A vicariance approach to historical biogeography. *Bioscience* 30, 339-343 (1980).

31. G. J. Nelson and N. I. Platnick. Systematics and Biogeography: Cladistics and Vicariance. Columbia University Press; New York, 567pp. (1981).

32. A. Nur and Z. Ben-Avraham. Lost Pacifica Continent: a mobilistic speculation. In: *Vicariance Biogeography: A Critique.* G. Nelson and D. E. Rosen (Eds.). pp. 341-358. Columbia University Press, New York (1981).

33. E. C. Olson. Biological and physical factors in the dispersal of Permo-Carboniferous terrestrial vertebrates. In: *Historical Biogeography, Plate Tectonics and the Changing Environment.* J. Gray and A. J. Boucot (Eds.). pp. 227-238. Oregon State University Press, Corvallis, Oregon (1979).

34. C. Patterson. The distribution of Mesozoic freshwater fishes. *Mém. Mus. Natio. d'Hist. Nat. A, Zool.* 88, 156-174 (1975).

35. C. Patterson. Methods of paleobiogeography. In: *Vicariance Biogeography: A Critique.* G. Nelson and D. E. Rosen (Eds.). pp. 446-489. Columbia University Press, New York (1981).

36. C. Patterson and D. E. Rosen. Review of ichthyodectiform and other Mesozoic teleost fishes and the theory and practice of classifying fossils. *Bull. Amer. Mus. Nat. Hist.* 158, 81-172 (1977).

37. E. C. Pielou. Biogeography. John Wiley and Sons, New York (1979).
38. N. I. Platnick and G. J. Nelson. A method of analysis for historical biogeography. *Syst. Zool.* **27**, 1-16 (1978).
39. D. E. Rosen. A vicariance model of Caribbean biogeography. *Syst. Zool.* **24**, 431-464 (1975).
40. D. E. Rosen. Vicariant patterns and historical explanation in biogeography. *Syst. Zool.* **27**, 159-188 (1978).
41. M. Sanders. Die fossilen fische der alttertiaren süsswasserablagerungen aus Mittel-Sumatra. *Verhandel. Geol.-Mijnbouw. Genoot. Nederland en Kolonien (Geol. Ser.)* **11**, 1-144 (1934).
42. R. Da Silva Santos. Laeliichthys ancestralis, Novo Gênero e Espécie de Osteoglossiformes do Aptiano da Formaçao areado, estado de minas gerais, Brasil. MME-DNPM, *Geologia* **27**, Paleont. estrati. **2**, 161-167 (1985).
43. B. Schaeffer. A teleost from the Livingston Formation of Montana. *Amer. Mus. Novit.* **1427**, 1-16 (1949).
44. A. G. Smith and J. C. Briden. Mesozoic and Cenozoic Paleocontinental Maps. Cambridge Earth Science Series, Cambridge University Press, London (1977).
45. A. G. Smith, A. M. Hurley and J. C. Briden. Phanerozoic paleocontinental world maps, 102pp. Cambridge University Press, Cambridge (1981).
46. A. G. Smith, D. G. Smith and B. M. Funnell. Atlas of Mesozoic and Cenozoic Coastlines, 99pp. Cambridge University Press, Cambridge (1994).
47. Su De-zao. The discovery of a fossil osteoglossid fish in China. *Vert. PalAsia.* **24**, 10-19 (1986).
48. E. C. Sytchevskaya. Palaeogene freshwater fish fauna of the USSR and Mongolia. Trude Sovmestnaya Sovetsko-Mongol'skaya Paleontologicheskaya Ekspeditsiya **29**, 1-157 (Russian) (1986).
49. L. Taverne. Ostéologie, phylogénese et systématique des téléostéens fossiles et actuels du superordre des ostéoglossomorphes. *Troisieme Partie. Acad. Roy. Belg., Mém. Classe Sci.*, XLIII-Fascicule **3**, 1-168 (1979).
50. M. Waldman. Fish from the freshwater Lower Cretaceous of Victoria, Australia, with comments on the palaeo-environment. Special papers in Palaeontology, Number 9, published by the Palaeontological Association, London, 124pp. (1971).
51. A. Wegener. The Origin of Continents and Oceans. Translated by John Biran in 1966, New York: Dover Publications (1929).
52. E. O. Wiley. Parsimony analysis and vicariance biogeography. *Syst. Zool.* **37**, 291-290 (1988).
53. M. V. H. Wilson and R. R. G. Williams. Phylogenetic, biogeographic, and ecological significance of early fossil records of North American freshwater teleostean fishes. In: *Systematics, Historical Ecology, and North American Freshwater fishes.* R. L. Mayden (Ed.). pp. 224-244. Stanford University Press, Stanford, California (1993).
54. Zhang (=Chang) M. -m and Zhou (=Chou) J.-J.. Discovery of *Plesiolycoptera* in Songhuajiang-Liaoning Basin and origin of Osteoglossomorpha. *Verte. PalAsia.* **15**, 146-153 (1976).

Proc. 30th Int'l. Geol. Congr., Vol. 12, pp. 67-78
Jin and Dineley (Eds)
© VSP 1997

Sequence of Events around the K/T Boundary at El Kef (NW Tunisia)

PIERRE DONZE[1], HENRIETTE MÉON[1], ERIC ROBIN[2], ROBERT ROCCHIA[2],
OUMRANE BEN ABDELKADER[3], HABIB BEN SALEM[3], and ANNE-LOUISE
MAAMOURI[3]

[1] *Université Claude-Bernard Lyon 1, Centre de Paléontologie stratigraphique et Paléoécologie,*
UMR 5565 CNRS, 29 bd du 11 novembre 1918, F 69622, Villeurbanne France.
[2] *Centre des faibles radioactivités, Laboratoire mixte CEA-CNRS, avenue de la Terrasse. F*
91198 GIF-sur-Yvette Cedex, France.
[3] *Office national des Mines, 24 rue 8601, 2035 La Charguia, Tunisie.*

Abstract

The section at El Kef (NW Tunisia) has been selected as the stratotype of the K/T boundary. It offers exceptional conditions for a very accurate study of the biological crisis. Three events can be identified from geochimical and paleontological studies. Two of them developed over long periods of time: a marine regression and a climatic cooling. The third one is a cosmic catastrophe characterized by the presence of cosmic material (Iridium, Ni-rich spinels...) in a thin millimetric brown-reddish layer (goethite). This layer is now considered as the K/T boundary. The marine regression produced gradual changes in the planktonic assemblages and favoured the development of taxa adapted for shallow waters. The climatic change modified the composition of terrestrial supply (spores and pollen grains); tropical taxa gradually decrease in number till total extinction and are replaced by taxa with more European affinities. The cosmic event produced the most dramatic effects: immediately above the goethite layer, 90 to 95 % of marine populations with carbonate tests are missing. The remaining 5 to 10 % get progressively extinct in the above 10 cm marly sediments. Ante-crisis conditions are recovered about two meters above the boundary. On the contrary, dinoflagellates, algae with a chitinous cyst, seem much less affected by the K-T event. Continental palynoflora are also seriously disturbed by the K-T boundary: in the first centimeters above the boundary their rate of extinction is about hundred times higher than observed in the upper Maastrichtian. However, many taxa survived. The most widely accepted hypothesis is that the large quantity of dust produced by the cosmic collision resulted in the absorption of solar radiations and the reduction of photosynthesis. However, this very attractive hypothesis does not fit El Kef observations: the absorption of solar light should have also dramatically affected dinoflagellates. This is not what we observe. Data suggest that, in addition to the reduction of photosynthesis, the cosmic event produced chemical conditions in the ocean which disturbed the formation process of calcareous tests.

Keywords: K/T boundary, El Kef stratotype section, Tunisia, cosmic event, marine regression, climatic change

INTRODUCTION

The region of El Kef (NW Tunisia) is a well known site for research about the K/T boundary formations. As early as the beginning of this century, the eminent geologist Pervinquière [23] suggested that a continuous transition between Cretaceous and Tertiary existed in that area. This was confirmed by subsequent studies and consequently this site became a classical place for biostratigraphy of the Maastrichtian and the Paleocene (6th and 7th Colloquium of African Micropaleontology, Tunis 1974, Ile-Ife 1976 and 26th IGC, Paris), [8, 2].

The importance of this site is due to its peculiar paleogeographical position. It was situated on the southern edge of the Tethys which extended southward over the large Saharan platform. During the late Cretaceous central Tunisia was affected by some positive shifts of the platform, which, with the regression of the sea, resulted in the formation of a large, eroded, continental hinterland in the South of Tethyan sea ("Ile de Kassérine"). The El Kef region at that time occupied a position between the outer continental plateau and the edge of the inner basin. In this clearly open but moderately deep area (ca. 200-300 m), pelagic and benthic organisms found good conditions for development, whereas, with a fairly flat region in the hinterland, no coarse detritus disturbed the sedimentation.

THE EL KEF SECTION (Fig. 1-2)

Several extended outcrops of Maastrichtian - Paleocene layers are visible SW of El Kef (geological map El Kef, 1/50000 [6])] near the dust road which, going from Marabout Si Abd Allah Srhir, on the Tajerouine road, joins the village of Hamman Mellègue (Fig. 1).

Above the calcareous lower Maastrichtian, a 500m thick marly series, the El Haria Formation shows continuous sedimentation corresponding to the upper Maastrichtian and Paleocene. The K/T boundary crops out near the hamlet of Fedj Hajar (Lambert coord.: X = 387,6; Y = 317,65).

The uppermost Maastrichtian, the top of the Mayaroensis zone (Fig.2, unit A), consists of grey compact marls with 40-50 % $CaCO_3$. On top of this sequence a thin, rusty-brown, goethite layer (unit B: 1-2 mm) is intercalated between two thin gypsiferous layers. In spite of its thinness, this goethite layer is continuous and can be followed over several hundred meters. This layer is the reference level for sampling. It is overlain by a dark clayey marl (unit C) and a black plastic clay layer (D : about 3 cm) without $CaCO_3$. Above this, the carbonate fraction (unit E) increases slowly and reaches 50%, as in the upper Maastrichtian, about two meters above the reference level.

INTRA-MAASTRICHTIAN EVENTS

The upper Maastrichtian contains abundant fossils of planktonic and benthic microorganisms : foraminifers [2, 29,3, 13, 15, 14], ostracods [4], coccoliths [21,22,24] and dinocysts [5]. There are also spores and pollen grains from the emerged hinterland. The low bioturbation makes possible a very precise stratigraphic study. Quiet conditions of sedimentation produced good fossil preservation.

LEGEND

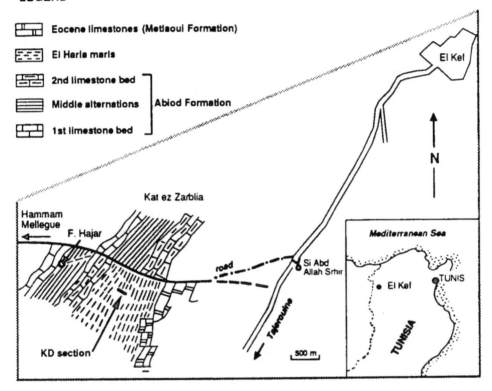

Figure 1. Location map of the El Kef section, NW Tunisia.

Marine Regression

The foraminifer assemblages remain abundant till the top of the Maastrichtian but their distribution shows an evolution in the upper part of the mayaroensis zone, i.e in the last 10 m of the Maastrichtian. The clear decrease of populations of Globotruncanidae, which require a minimum water depth for their development (at least 100m), to the benefit of Hete-rohelicidae more adapted to shallower waters, is the unquestionable indication of a marine regression.

Dinoflagellates studies support that view: Brinkhuis and Zachariasse [5] report the evidence of a regressive trend in the upper Maastrichtian.

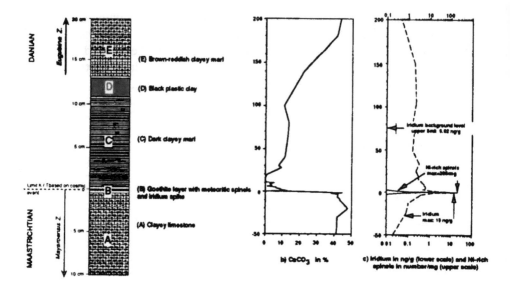

Figure 2. Cretaceous-Tertiary boundary in the El Kef section.
From the left to the right:- stratigraphical log near the boundary from -10 cm below the goethite layer (0) to +20 cm above it;- curve of CaCO3 content (from -50 cm below the goethite layer to +200 cm);- Iridium and Ni-rich spinel concentrations (from -50 cm to +200 cm); the lower scale corresponds to Ir ng/g and the upper scale to number of spinels/mg.

Climatic Cooling

The Continental Vegetation

The El Kef section was studied from a palynological point of view since the Falsostuarti zone [18] (165 m below the K-T boundary) to the Pseudobulloides zone (45 m above). The large number of described taxa (213) permits a very precisely study of the evolution of the terrestrial vegetation.

The continental palynoflora exhibits gradual changes over the late Maastrichtian. The rate of extinction is rather low (45 species over 45 meters of the uppermost Maastrichtian) and results essentially in the progressive desappearance of taxa of the Afro-South American paleotropical province (17-19). Most part of those which developed in the basal Danian are of european affinity. These gradual changes are clear indications of a climatic cooling possibly related to the marine regression and/or the enhanced volcanic activity which produced at that time the basaltic traps in Deccan [7].

The Marine Realm

The foraminifer and isotope studies [15] also suggest a gradual climatic cooling during late Cretaceous.

The cooling is also recorded by ostracods with the gradual disappearance of thermophilous forms (*Cytherelloidea* for instance, [8]).

K-T EVENT

Cosmic Markers

The K-T boundary at El Kef is marked, as in most K-T sections, by an overabundance of iridium and the presence of crystals of Ni-rich spinel, a mineral resulting from the oxidation of meteoritical material [27] The association iridium-Ni-rich spinel shows that a catastrophic cosmic event did occur at this time [26-28]. Iridium is spread over several meters due to post depositional diffusion and/or to the residence time of this metal in the ocean. In contrast, Ni-rich spinel is found only in the thin millimeter-thick brownish goethite layer at the very top of the Maastrichtian. Such a narrow distribution shows that the infall of cosmic matter resulted from a single hexplosive event [26].

Marine Organisms

The evolution of rich and highly diversified populations of calcareous marine micro-organisms is abruptly stopped right at the goethite layer. Foraminifers show a major reduction of assemblages immediately above it: nearly 90 % of the populations disappear. In the next 5 centimeters of sediments, pelagic forms get less and less abundant and disappear whereas the benthic forms survive with a selection of more tolerant forms (*Lenticulina, Cibicides, Ammodiscus*). Foraminifer content reaches a minimum in the black clay layer (unit D) which contains only very rare, small species (*Bulimina*, Agglutinants) more tolerant to anoxic conditions. We observe not only a drastic reduction of individual abundances but also a significant change in the aspect of specimens. While, in the uppermost Mastrichtian sizes are normal and tests are well calcified, above the goethite layer specimens are small and poorly calcified, giving evidence of very unfavourable conditions.

The assemblages of calcareous nannofossil [21, 22, 24] show great abundance and diversity without significant change up to the top of the Maastrichtian; but their frequency decreases drastically above the goethite layer and they are totally absent in the black clay layer (the samples LMC 6 to 8 in Perch-Nielsen [21] are equivalent to what we now call the goethite layer and the black clay layer). In addition, Perch-Nielsen notes that *Thoracosphaera*, a calcareous dinoflagellate cyst, is absent in the same levels.

Ostracods, benthic metazoa, which are very sensitive to the environment, were also much affected by the crisis. Some species (*Krithe* sp., *Martinicythere* cf. *vesiculosa* Apostolescu) are present in the first centimeter above the goethite layer but disappear in the next few centimeters [9, 11].

The dinoflagellates have already been studied by Brinkhuis and Zachariasse [5] in the Kef section. We studied again these organisms in order to evaluate their evolution in the immediate vicinity of the K/T boundary (Fig.3). They show no accelerated rates of extinction across the boundary. We observe only a decrease in their relative abundance (with respect to all palynomorphs) starting just below the goethite layer. We do not know if this fact is the consequence of the K-T event or due to a regressive pulse at the end of the Maastrichtian.

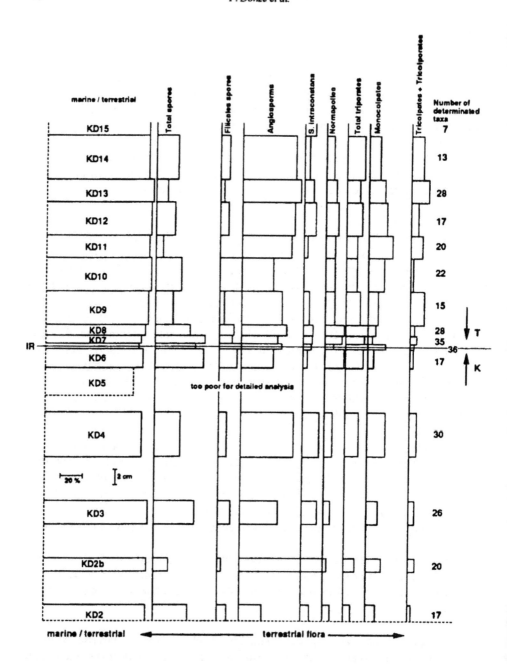

Figure 3. Palynological diagram of the El Kef section.
The thickness of each bar of the histogram corresponds to the thickness of each analysed sample.
The first left column presents the distribution of marine palynomorphs versus continental ones. It
shows the regressive trend of the sea in the upper Maastrichtian which culminates at the K/T
boundary and the transgression above it, always in an open marine environment.

The next columns show the distribution of main groups of the continental palynomorphs. For the Filicales spores, there is a maximum at the K/T boundary (=IR); the fern content goes from 15 to 20 % and thus is very different from the fern spike of USA: 100%.

The last right column gives the number of determinated taxa for each sample in the KD section, it can be seen that this number increases at the boundary and in the 2 next centimeters. This increasing corresponds more to an increasing continental supply (due to the lowering sea level) than to other facts but prooves that there is not a strong devastation of terrestrial flora, even if the disappearances of taxa increase at the boundary (see next figure).

Continental Vegetation

The continental contribution in sporomorphs increases at the K/T boundary level (factor 3 according to our data or even 4 to 5 according to the results of Brinkhuis and Zachariasse [5].

The spore/pollen assemblages are rich with a maximum diversity in the goethite layer and immediately overlying levels (Fig.4). The rate of extinction of taxa increases much in the goethite layer: 11 taxa disappear right in coincidence with the brief Ni-rich spinel infall. 12 additional taxa disappear in the next 20 cm till the top of the black plastic clay. This is an extinction rate a hundred times higher than in the upper Maastrichtian. However we cannot say that there is a collapse of the vegetation: about 50 taxa (among the main ones), i.e. 2 thirds of the sporomorphs, survived the boundary event. Obviously, in the Kef region the continental vegetation is far less affected than in North America where a devastation of terrestrial vegetal ecosystems is observed [25]. The same conclusion is derived from ferns: we observe at El Kef a small fern maximum (Fig.3) but nothing comparable with the enormous fern spike observed in north american continental sections.

Discussion

It is clear that the rapid collapse of the marine productivity at the K-T boundary (foraminifers, ostracods, coccolithophores...) occurred in the goethite layer, in coincidence with the sudden accretion of cosmic material resulting from the collision with an extraterrestrial body [1, 26-29]. Supposing that the clay fraction accumulation rate remained constant across the boundary (about 5 cm/1,000yrs), we can estimate that the biological crisis developped in less than 100 years (corresponding to the thickness of the goethite layer) and that the duration of the period of stress lasted at least 2,000 years (the top of the black plastic clay). Paleontological and geochemical analyses show that the marine environment became suddenly highly reducing [6, 15]. This lack of oxygen is almost certainly due to the low productivity of phytoplankton (coccolithophores) which are dependant on solar radiations. It has been proposed that a dust blanket resulting from the cosmic collision produced a drastic absorption of solar radiations and, consequently, a severe reduction of photosynthesis. This scenario is highly attractive. However, it cannot account for El Kef observations.

The main objection is that it does not explain the differences of behaviour between calcareous nannofossils and calcareous dinoflagellates (*Thoracosphaera*) , on one hand, and chitinous dinoflagellates, on the other hand. They all need solar light for their development but, surprisingly, the former completely disappeared while most of the

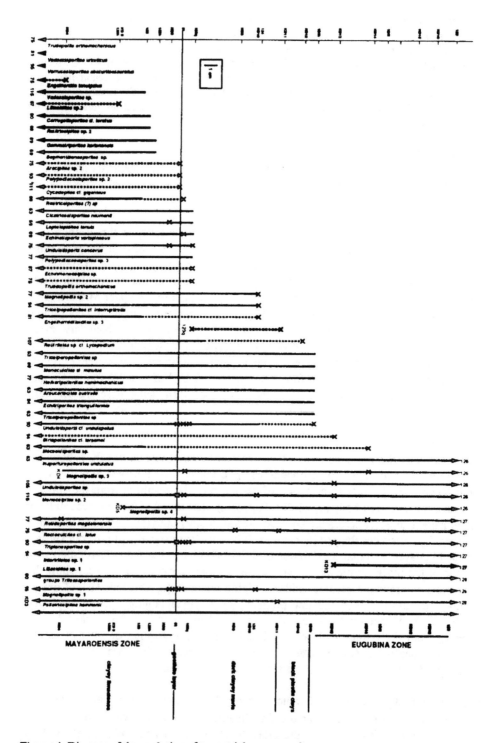

Figure 4. Diagram of the evolution of terrestrial sporomorphs.

In this diagram, samples labeled 122b to 126 were collected for our first publication [17], KD 4 to 19 samples provided by a new sampling (left column).

Because of drawing conveniences, it was not possible to give the evolution on the whole Mayaroensis zone, 45 species disappear in the last 45 m of upper Maastrichtian. For the same reason, all the taxa remaining above the black plastic clays are not represented, there are about 50 taxa.

The lower numbers indicate the appearance level of each species (for instance, 63, the lowermost sample belongs to the Falsostuarti zone, more than 100 m below the boundary).

Most of taxa passing the black plastic layer continue till the Pseudobulloides zone, about 30 m above the boundary.

later survived. Obviously the amount of solar light is not the sole parameter driving the extinction pattern.

A second objection is that, if the continental vegetation recorded at El Kef were seriously disturbed by the K-T event (extinctions of 1/3 of the species), it is not totally destroyed and replaced by ferns as observed in North-american continental sites [20, 25, 31]. What we observe at El Kef is a moderate disturbance of the vegetation on the continent and dinoflagellate in the ocean which contrast with the extremely severe destruction of calacareous planktic species. Such a striking selectivity could be explained by the influence of an additional factor.

We are pushed to hypothesize that the severity of the crisis for calcareous marine species was amplified by the difficulty of microorganisms to produce carbonate. It is well known that the pH and the abundance of trace elements play a fundamental role in the development of microorganisms, particularly in the construction of their carbonate shells and tests. Two processes can be envisaged: 1) acid rains [16] resulting from the massive formation of nitrogen oxides in the atmosphere and from the injection of sulfur oxides derived from the dissociation of the target material; 2) a contamination of the ocean surface water by heavy metals deriving from the impactor and ionized in sea water [3]. The resulting change of the ocean surface conditions could have precluded the production of carbonate by coccolithophores. Their disappearance would have thus deprived the marine environment of oxygen which the animal realm relies on. This could account for the disappearance of the microfauna and nannoplankton and, consequently, the sudden passage from carbonate to clay sedimentation as it can be seen everywhere in the low and mid latitudes.

CONCLUSION

Three different and important events are recorded in El Kef section: a marine regression, a global cooling and a cosmic catastrophe. The marine regression is illustrated by the decrease in abundance of organisms which require a minimum water depth for their development to the benefit of populations more adapted to surficial waters. The climatic cooling is clearly visible in the record of continental palynoflora which exhibits gradual changes. The marine regression and the climatic cooling are long duration and slowly evolving events showing their effects well before the end of the Maastrichtian (Mayaroensis zone). They are probably related to the same cause, possibly the instability of the lithosphere inducing a strong tectonic and volcanic activity [7] and, hence, a cooling of the atmosphere. The third event is a cosmic catastrophe. It is recorded in the thin layer of goethite which contains iridium and Ni-rich spinel crystals and marks a drastic reduction of calcareous microfossils. The temporarily surviving faunas, accepting strongly anoxic conditions, disappear completely about 10 centimeters above the goethite layer, in a black plastic clay layer. The stratigraphic distribution of the Ni-rich spinel crystals at El Kef is very narrow (a few mm) and coincide exactly with the sheer drop of the calcareous microfossils, demonstrating the causal link between the biological crisis and the cosmic catastrophe. The consequences of this catastrophe are well visible in the marine realm and to a lesser extent in the continental domain. It is difficult to explain the selectivity of the extinctions by the absorption of solar radiations only. We suggest that the importance of the crisis in the marine realm was increased by chemical conditions in sea water preventing the development of species with calcareous tests.

REFERENCES

1. L.W. Alvarez, W. Alvarez, F. Asaro and H.Y. Michel H.Y. Extraterrestrial cause for the Cretaceous-Tertiary extinction. *Science*, 208, 1095-1108 (1980).
2. J.P. Bellier, M. Caron, P. Donze, D. Herm, A.L.Maamouri and J. Salaj. Le Campanien sommital et le Maastrichtien de la coupe du Kef (Tunisie septentrionale) : zonation sur la base des Foraminifères planctoniques, *Zitteliana*, 10, 609-611 (1983).
3. O.B. Abdelkader. Planktonic foraminifera content of El Kef Cretaceous-Tertiary (K/T) boundary type-section (Tunisia), *Workshop on Cretaceous-Tertiary transitions at El Kef. Abstracts*, 9 (1992).
4. O.B. Abdelkader, H.B. Salem, P. Donze, L. Froget, A.L. Maamouri, H. Méon, E. Robin and R. Rocchia. The K/T stratotype section of El Kef (Tunisia): events and biotic turnovers, *Geobios, Mem.sp.*(in press).
5. H. Brinkhuis and W.J. Zachariasse. Dinoflagellate cysts, seal level changes and planktonic foraminifers across the Cretaceous-Tertiary boundary at El Haria, northwest Tunisia, *Marine Micropaleontology*, 13, 153-191 (1988).
6. P.F. Burollet and P. Sainfeld. Notice explicative de la feuille au 1/50000 n° 44 Le Kef, 32 (1956).
7. V. Courtillot. Deccan volcanism at the Cretaceous - Tertiary boundary : post climate crisis as a key to the future ? , *Palaeogeog., Palaeoclimat., Palaeoecol. (Global and Planetary Change Sect.)*, 189, 291-299 (1990).
8. P. Donze. Une série de référence pour le Maastrichtien et le Paléocène en faciès mésogéen : la coupe dite de la "piste du Hamman Mellègue" au SW du Kef (Tunisie septentrionale, *26e C.G.I.*, (Paris 1980), 1, 225 (1980).
9. P. Donze. Evolution of the Ostracods microfauna at the Cretaceous-Tertiary boundary in El Kef section (NW Tunisia), *Workshop on Cretaceous-Tertiary transitions at El Kef. Abstracts*, 2-3 (1992).
10. P. Donze, S. Jardiné, O. Legoux, E. Masure and H. Méon. Les évènements à la limite Crétacé-Tertiaire: au Kef (Tunisie septentrionale), l'analyse palynoplanctologique montre qu'un changement climatique est décelable à la base du Danien, *Actes 1er Cong. Nat. Sciences Terre*, Tunis, 161-169 (1981).

11. P. Donze, J.P. Colin, R. Damotte, H.J. Oertli, J.P. Peypouquet and R. Saïd. Les Ostracodes du Campanien terminal à l'Eocène inférieur de la coupe du Kef, Tunisie nord-occidentale, *Bull. Centres Rech. Explor-Prod. Elf-Aquitaine*, **6**, 2,. 273-335 (1982).

12. P. Donze, H. Méon, R. Rocchia, E. Robin and L. Froget. Biological changes at the KT stratotype of El Kef (Tunisia). *In* New developments regarding the KT event and other catastrophes in Earth history, *LPI Contribution n° 825, Geol. Soc. Amer., Sp. pap.*, **247**, 72-76 (1994).

13. G. Keller. Extinction, survivorship and evolution of planktic foraminifera accross the Cretaceous/Tertiary boundary at El Kef, Tunisia, *Marine Micropaleontology*, **13**, 239-263 (1988).

14. G. Keller, L. Li and N. MacLeod. The Cretaceous / Tertiary boundary stratotype section at EL Kef, Tunisia: how catastrophic was the mass extinction ? , *Palaeogeog., Palaeoclimat., Palaeoecol.*, **119**, 221-254 (1995).

15. G. Keller and M. Lindinger. Stable isotope, TOC and Ca CO_3 record across the Cretaceous/Tertiary boundary at El Kef, Tunisia, *Palaeogeog., Palaeoclimat., Palaeoecol.*, **73**, 243-265 (1989).

16. D.M. McLean. Deccan traps mantle degassing in the terminal Cretaceous marine extinctions, *Cretaceous Research*, **6**, 235-259 (1985).

17. H. Méon. Palynologic studies of the Cretaceous-Tertiary boundary interval at El Kef outcrop, northwestern Tunisia : Paleographic implications. *Rev. Palaeobot. Palynol.*, **65**, 85-94 (1990).

18. H. Méon. Etudes sporopolliniques à la limite Crétacé-Tertiaire : la coupe du Kef (Tunisie nord-occidentale) ; étude systématique, stratigraphie, paléogéographie et évolution climatique. *Palaeontographica*, B, **223**, 107-168 (1991).

19. H. Méon and P. Donze. Etude palynologique du passage Crétacé-Tertiaire dans la région du Kef (Tunisie NW). L'environnement végétal terrestre et son évolution, *Inst.fr. Pondichery, Trav. sect. sci. techn.*,**25**, 237-350 (1988).

20. D.J. Nichols. Geologic and biostratigraphic framework of the non-marine Cretaceous-Tertiary boundary interval in Western North America, *Rev. Palaeobot. Palynol.*, **65**, 75-84 (1990).

21. K. Perch-Nielsen. Nouvelles observations sur les Nannofossiles calcaires à la limite Crétacé-Tertiaire près de El Kef (Tunisie). *Cahiers Micropal.*, **3**, 25-36 (1981).

22. K. Perch-Nielsen, J. Mckenzie, H.E. Quziang, L.T. Silver and P.H. Schultz. Biostratigraphy and isotope stratigraphy and the "catastrophic" extinction of calcareous nannoplankton at the Cretaceous/Tertiary boundary, *Geol. Soc. Amer., Sp. Pap.* **190**, 353-371 (1982).

23. L.Pervinquière. 1903 - *Etude géologique de la Tunisie centrale.*. Thèse Doct. ès Sciences Univ. Paris : 359 p. (1903).

24. J.J. Pospichal. Calcareous nannofossils at the K-T boundary, El Kef : No evidence for stepwise, gradual, or sequential extinctions. *Geology*, **22**, 99-102 (1994).

25. C.J. Orth, J.S. Gilmore, J.D. Knight, C.L. Pillmore, R.H. Tschudy and J.E. Fassett. An Iridium abundance anomaly at the palynological Cretaceous-Tertiary boundary in northern New Mexico, *Science*, **214**, 1341-1343 (1981).

26. E. Robin, D. Boclet, P. Bonté, L. Froget, C. Jéhanno and R. Rocchia. The stratigraphic distribution of Ni-rich spinels in Cretaceous-Tertiary boundary rocks at El Kef (Tunisia), Caravaca (Spain) and Hole 761C (Leg 122), *Earth Planetary Science Letter*, **107**, 715-721 (1991).

27. E. Robin., Ph. Bonté, L. Froget, C. Jéhanno and R. Rocchia. Formation of spinels in cosmic objects during atmospheric entry: aclue to the Cretaceous-Tertiary boundary event. *Earth and Planetary Science Letter*, **108**, 181-190 (1992 a).

28. E. Robin, P. Bonté, P. Donze, L. Froget, C. Jehanno and R. Rocchia. The Ni-rich spinel distribution at K/T boundary of El Kef, Tunisia : Evidence for a short catastrophic cosmic event, *Workshop on Cretaceous-Tertiary transitions at El Kef - Abstracts*, 17 (1992b).

29. R. Rocchia, D. Boclet, P. Bonte, P. Donze, C. Jehanno, L. Froget and E. Robin. The K/T event time-scale and the Iridium anomaly. The importance of the site of El Kef, Tunisia, *Workshop on Cretaceous-Tertiary transitions at El Kef - Astracts*, 15-17 (1992) .

30. J. Smit, A.J. Nederbragt, W. Alvarez, A. Montanari and N. Swinburne. The Cretaceous-Tertiary (K/Pg) boundary type section of El Kef, Tunisia, compared with proximal (K/Pg) transition sections from the gulf of Mexico. *Workshop on Cretaceous-Tertiary transitions at El Kef-Abstracts*, 4-6 (1992).

31. R.H. Tschudy, C.L. Pillmore, C.J. Orth, J.S. Gilmore and J.D. Knight. Disruption of the terrestrial plant ecosystem at the Cretaceous-Tertiary boundary, western interior, *Science*, **225**, 1030-1032.

Proc. 30ᵗʰ Int'l. Geol. Congr., Vol. 12, pp. 79-94
Jin and Dineley (Eds)
© VSP 1997

Comparative Faunal Content of Strunian (Devonian) between Etaoucun (Guilin, Guangxi, South China) and the Stratotype Area (Etroeungt, Avesnois, North of France)

BRUNO MILHAU[1,5], BRUNO MISTIAEN[1,5], DENISE BRICE[1], JEAN MARIE DEGARDIN[2,5], CLAIRE DERYCKE[2,5], HOU HONGFEI[3], JEAN CLAUDE ROHART[1], DANIEL VACHARD[2,5] and WU XIANTAO[4]

[1] *Laboratoire de Paléontologie stratigraphique, F.L.S.-I.S.A., 13 rue de Toul, 59046 Lille Cedex. France.*

[2] *Laboratoire de Paléobotanique, U.F.R. Sciences de la Terre, U.S.T.L., 59655 Villeneuve d'Ascq. France.*

[3] *Institute of Geology, Chinese Academy of Geological Sciences, Baiwanzhuang Road, Beijing 100037 China.*

[4] *Jiaozuo Institute of Technology, Jiefang Road, Henan, 454159 China.*

[5] *URA 1365, CNRS.*

Abstract

In the Etaoucun area, near Guilin in Guangxi (South China), a continuous section is exposed through the Devonian-Carboniferous boundary, in a platform environment. The following formations have been sampled in detail: the upper part of the Dongcun Formation (Upper Famennian), the Etaoucun Formation (Uppermost Famennian = Strunian) and the lower part of the Yaoyunling Formation, Shangyueshan Member (Lower Carboniferous).

Foraminifera, stromatoporoids, tabulate and rugose corals, brachiopods, ostracods, conodonts and vertebrates micro-remains have been investigated and compared with Strunian faunas of the stratotype area (Avesnois, North of France) near Avesnes (Godin), Avesnelles and Etroeungt, where precise bed-to-bed sampling has been done.

Excepting foraminifera, which allow some correlations, the benthic fossil groups present very few taxa in common; this supports the endemism of the South-China Strunian faunas already emphasized by some authors.

Keywords: D/C Boundary, Famennian, Strunian, South China, Etroeungt, France, brachiopods, conodonts, foraminifera, vertebrates micro-remains, ostracods, rugose corals, stromatoporoids, tabulate corals.

INTRODUCTION

The Tangjiawan-Etaoucun section is located about 8 km S-SW of Guilin (Guangxi, South China) and about 15 km south of the D/C boundary parastratotype, near Nanbiancun (fig. 1). It displays a more than 1200 m thick continuous sequence of

Devonian-Carboniferous rocks deposited on a more or less restricted platform [2], with
reef facies, back reef facies and lagoon facies.

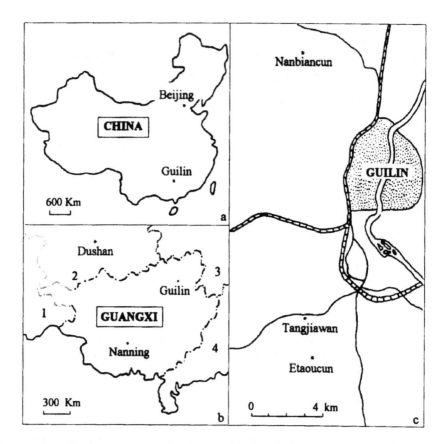

Figure 1. Location maps. 1a - Location of the studied area in South-China. 1b - Location of the
studied area in Guangxi Province (after Zhou [1]); 1: Yunnan, 2: Guizhou, 3: Hunan, 4:
Guangdong. 1c - Location of the section in the South/South-West of Guilin (after Yu [2]).

Near the village of Etaoucun, a well developed section was described [3], and published
later with slight differences [4]. Three formations can be recognized (table 1).

The Dongcun Formation (= "Rongxian" Formation), about 495 m thick, of Famennian
age, consists of light grey, thick, bird eye , usually unfossiliferous limestones. In the
lowermost part of the Formation few *Amphipora* have been discovered [5], and in the
uppermost part several Leperditiidae ostracod-bearing beds have been observed [6], and
one reefal level, with stromatoporoids, recognized for the first time.

The Etaoucun Formation, about 80 m thick, of Uppermost Famennian age (= Strunian
s.l.), consists of grey and dark thick bedded limestones; numerous reefal levels are
present.

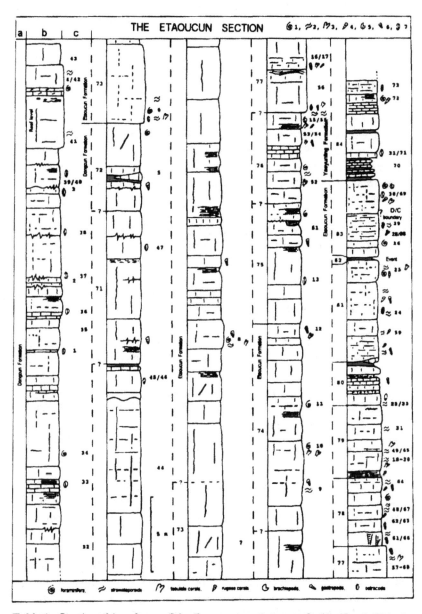

Table 1. Stratigraphic column of the Etaoucun section near Guilin (South China). a: Chinese unit numbers [4]; b: Lithostratigraphic column; c: Sampling numbers in the three different recognized formations and significant fossiliferous levels. 1: foraminifera; 2: stromatoporoids; 3: tabulate corals; 4: rugose corals; 5: brachiopods; 6: gastropods; 7: ostracods.

The Yaoyunling Formation, Shangyueshan Member, of Lower Carboniferous age, mainly consists of thin bedded, dark limestones, containing a more open-water marine fauna.

STRATIGRAPHY

The Etaoucun Section
The Etaoucun section was sampled in detail by B. Milhau and B. Mistiaen in August 1992 and January 1995 (21 samples from the upper part of the Dongcun Formation, 45 in the Etaoucun Formation and 7 in the first ten meters of the Yaoyunling Formation, Shangyueshan Member). The stratigraphic column is given in table 1 and prompts the following three remarks:

1)The total thickness we measured for the Etaoucun Formation is over 100 meters; this is in accordance with Hou's statement [7] but in conflict with published descriptions [3, 4] which only give 75,5 m.

2)The units 75 to 84 of the Chinese authors are recognized, with similar thicknesses, except for the units 80 + 81 where we have 10 m against 8,60 m [3] or 20,50 m [4].

3)The lower part of the unit 73 is also well recognized, but we measured 59 m for beds 73 + 74 instead of the published 25 m; no fault has been observed in the field.

The Strunian of the Stratotype Area
In 1857, Gosselet [8] gave the first description of a calcareous-shaly sequence observed in the Le Parcq quarry, East of Etroeungt, in Avesnois, Northern France (fig. 2); he formally designated this "Calcaire d'Etroeungt" in 1860 [9], and the section will be proposed as the stratotype of the Strunian [10, 11].

A precise bed-to-bed sampling of the strata in this quarry allowed recognition of most of the beds described by Gosselet, except his "dernier banc" at the base. Then, in the Le Parcq quarry, the sequence is complete, with all due respect to Mamet *et al.* [12], and can be fully observed without any gap; this can be demonstrated with a section in the nearby Jean-Pierre quarry.
Nevertheless, many works, especially those of Conil *et al.* [11, 13, 14, 15, 16], have shown the restricted place occupied by the "Calcaire d'Etroeungt" in the transgressive uppermost Devonian sequence. This can be observed further in the north (fig. 2) in the "tranchée d'Avesnelles" railway cut, SE of Avesnes and in the "St Hilaire" section in the NW.

Then the idea of a Strunian stage s.l. was introduced [13], mainly for reasons essentially based on Foraminifera data; the parastratotype was proposed in the "tranchée d'Avesnelles" [11, 15, 17] where three units can be distinguished (table 2): a shaly Strunian, a sandy Strunian and a calcareous Strunian; the "Calcaire d'Etroeungt" s.s. (Gosselet) only represents the middle part of the "Calcaire d'Etroeungt" s.l. (Conil &

Lys), which is the calcareous Strunian.

Recently a new cut in the Bocahut quarry, near Godin (fig. 2), SW of Avesnes, allowed us to study a new interesting section of the Devonian/Carboniferous strata.

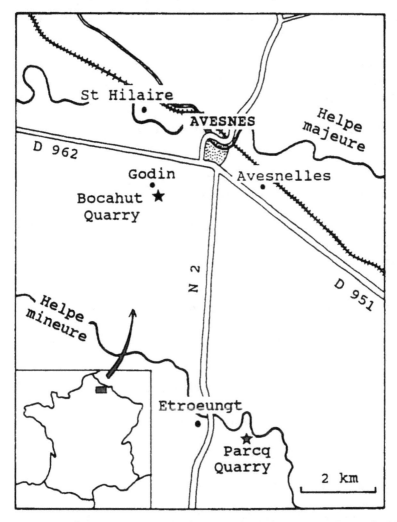

Figure 2. Location of the main Strunian outcrops in the stratotype area (Avesnois, North of France)

PALAEONTOLOGICAL CONTENT OF THE ETAOUCUN SECTION AND COMPARISON WITH THE STRATOTYPE AREA

Foraminifera (D. Vachard)
The distribution of foraminifera in the Etaoucun section was presented by Conil [4]. The following are the new data revealed:

1) the first appearance of *Septatournayella rauserae* LIPINA, 1955 and *Eoendothyra communis communis* (RAUSER, 1948) in the lower part of the sequence in sample CH-ET 34;

2) the first appearance of *Eoendothyra communis regularis* (LIPINA, 1955) and *Septaglomospiranella primaevae* (RAUSER, 1948) in sample CH-ET 6;

3) the first appearance of *Quasiendothyra kobeitusana* (RAUSER, 1948) and *Q. konensis* (LEBEDEVA, 1956) in sample CH-ET 11.

Besides species already reported by Conil, other associated foraminifera are *Eoendothyra communis delicata* DURKINA, 1959, *Avesnella* cf. *elegantula* (REITLINGER, 1961), *Baelenia* sp. and *Tournayella* sp.

In addition, we have the classic division into four biozones found in South China [19, 20, 21]:

1) the *Septatournayella rauserae rauserae* and *Eoendothyra communis communis* zone, in sample CH-ET 34, which corresponds to the Df3β of the Franco-Belgian stratotypes [22];

2) the *Septatournayella rauserae rauserae* and *Eoendothyra communis regularis* zone, from sample CH-ET 6, which corresponds to the Df3γ/δ;

3) the *Quasiendothyra kobeitusana* and *Q. konensis* zone, from sample CH-ET 11 and rising above an anoxic "event" located about 4 m below the top of the Etaoucun Formation, which is equivalent to the Franco-Belgian Df3ε and Cf1α;

4) the *Bisphaera* and *Earlandia* zone, probably starting from sample CH-ET 68, which corresponds to the Cf1α' (Lower Hastarian p.p.) of global distribution, so that an Hastarian Pangea has been proposed to account for the paleobiogeography of Foraminifera [23].

Besides foraminifera, the main associated calcareous microfossils are:

1) Solenoporacea algae with *Parachaetetes johnsoni* MASLOV, 1962;

2) Moravamminidae pseudo-algae including *Subkamaena razdolnica* BERCHENKO, 1981, *Devonoscala tatarstanica* (ANTROPOV, 1959), *Pseudonanopora* sp. and *Eouraloporella* sp.;

3) Incertae sedis including *Menselina clathrata* ANTROPOV, 1967.

Compared with the calcareous beds of the Bocahut and Le Parcq quarries, the microflora and microfauna carbonate associations are very similar, with only little differences. In both quarries, the Df3ε-Cf1α associations are characterized by the abundance of *Quasiendothyra kobeitusana*, *Endothyra parakosvensis* LIPINA, 1955 and *Paracaligelloides florennensis* (CONIL & LYS, 1964), which are present but rare in China. Moreover, pseudo-algae are more diversified in the Franco-Belgian basin with *Kettnerammina grandis* (CHUVASHOV, 1965), *Palaeoberesella* sp., *Kamaena* sp., *Subkamaena* sp., etc.

Stromatoporoids (B. Mistiaen)

Stromatoporoids are abundant in the Etaoucun section. Li *et al.* [4] pointed out their presence in their units 73, 76, 77, 79, 81 and 83 (Etaoucun Formation), with the following taxa: *Actinostroma* sp., *Anostylostroma* sp., *Pennastroma* sp., *Platiferostroma* sp., and *Stromatocerium* (= *Platiferostroma*) sp.

More than eighty stromatoporoid samples were collected from about fifteen different levels of the Etaoucun section. We have recognized the following taxa: *Trupetostroma* represented by two species, *Platiferostroma* also with two species and *Stylostroma* (= *Pennastroma* [24]) with three species.

Relative to the presence of stromatoporoids in this section, the following three points must be underlined.

1) We have located, about 35 m below the Etaoucun Formation, in the Dongcun Formation, a 5 m thick reefal level with stromatoporoids represented only by the genus *Platiferostroma*. Referring to foraminifera, this level belongs to the upper part of the *Septatournayella rauserae rauserae / Eoendothyra communis communis* (Df3β) biozone or to the lower part of the *Septatournayella rauserae rauserae / Eoendothyra communis regularis* (Df3γ/δ) biozone, and could correspond to the Upper Famennian (not yet Strunian).

2) We found, in samples CH-ET 29, located below the postullated D/C boundary [4], but still within the Etaoucun Formation, several specimens of *Trupetostroma* sp. A. We can observe in thin sections that they are associated with the foraminifera *Quasiendothyra* and/or *Earlandia* and *Bisphaera*. Then, from the foraminifera, those beds could correspond either to the top of the *Quasiendothyra kobeitusana / Q. konensis* biozone (Df3ε-Cf1α) or to the base of the *Bisphaera / Earlandia* biozone (Cf1α', Lower Hastarian). Unless the D/C boundary is not well-defined (see conclusion) in the section, then stromatoporoids could be Hastarian in age and could have survived, in this area, in the lowermost part of the Carboniferous. Nevertheless, those stromatoporoids could be reworked specimens from some lower levels; indeed, the species *Trupetostroma* sp. A is also present in sample CH-ET 25, located about 2.20 m beneath, that is to say below the anoxic "event".

3) The stromatoporoid fauna from the Etaoucun section (in the Dongcun and Etaoucun Formations) typically belongs to the Strunian "mixed assemblage", with labechiids and non labechiids (= assemblage 2), of Stearn [25] and Cockbain [26]; more precisely, it belongs to the assemblage 2a, characterized by genera such as *Platiferostroma* [27].

Stromatoporoids are also very common in the Le Parcq quarry (Avesnois), the stratotype of the "Calcaire d'Etroeungt". They have been previously studied and illustrated by Le Maître [28] who recognized eleven species; however, systematic revision is necessary and Stearn *et al.* [27] gave some new generic assigments. We have also collected about eighty specimens from 42 different levels, and recognized twelve different taxa.

Relative to the Strunian stromatoporoids from Avesnois, the following points seem interesting to underline.

1) No genus is common with the Etaoucun Formation.

2) As far as we know, the genus *Amphipora* has never been pointed out previously in the Upper Famennian; it has been found in about twenty beds of the Le Parcq quarry section [5]. The *Amphipora* branches are never very abundant but several sections are sometimes observable in the same thin sections. *Amphipora* is also present in the equivalent levels of the Bocahut quarry (7 km North of Etroeungt). Nevertheless, the genus *Amphipora* is totally absent from the Etaoucun Formation in South China [29].

3) The stromatoporoid fauna from Etroeungt belongs typically to the "non labechiid assemblage" [25, 26, 27]. Then, relatively to the distribution of stromatoporoids, largely differentiated provinces are present during Strunian time [29].

Tabulate Corals (B. Mistiaen)
In the two studied sections of Etaoucun (South China) and Le Parcq quarry (Avesnois), tabulate corals are essentially represented by syringoporids.

Nevertheless, in South China, they are largely more abundant and diversified, with probably four or five species which seem similar to the syringoporid fauna reported by Tourneur [21] in Hunan.

In the Le Parcq quarry, complete colonies of syringoporids are scarce and, more often, only represented by branch fragments in thin sections. But in the Bocahut quarry, about seven kilometers in the north, syringoporid colonies are more common. Nevertheless, syringoporids in Avesnois are probably not so diversified as in South China.

In Avesnois, we also have to emphasize the association of some tabulate corals with stromatoporoids which do not seem to be true *Caunopora* tubes (syringoporids), but more probably auloporids.

Lastly, in the Le Parcq quarry section, few samples of massive cerioid Tabulate corals have been found, and are referred to the genus *Yavorskia* (Cleistoporidae).

Rugose Corals (J.C. Rohart)
In the Etaoucun Formation of the Etaoucun section, *Cystophrentis* has been found in the Chinese units 78 and 81 [4]. From field work, the presence of solitary rugose corals is reported, in the Etaoucun Formation, in unit 76 (sample CH-ET 15), in unit 77, in unit 78, at the base of unit 79, in unit 81 and at the top of unit 83. Specimens studied in laboratory (sample CH-ET 15) are assigned to *Cystophrentis* sp. In the lower part of the Carboniferous Yaoyunling Formation, other fragments (sample CH-ET 72, unit 84, = *Cystophrentis ?*) have also been found. The genus *Cystophrentis* is characteristic of the *Cystophrentis* Zone [30] and was considered by Yu as equivalent to the *Zaphrentis* Zone [31] (base of Carboniferous). Now this zone is given as Uppermost Famennian, that is to say "Strunian" [32, 33]. The contrast is high between this Etaoucun fauna, very poor

in individuals and species, and the Nanbiancun fauna of the same age. This locality yielded corals distributed commonly in the whole section and assigned to 23 species and 11 genera [34]. The Etaoucun stromatoporoid-rich facies must be considered as shallower and less quiet than the Nanbiancun one, and consequently, less favourable for rugose corals.

In the type locality of the Etroeungt limestone, Le Parcq quarry (Avesnois), solitary rugose corals are not found very frequently in the shales or in the limestone beds. Richer outcrops exist in the area around Etroeungt (Avesnes, Avesnelles, etc...). Usually, corallites were accumulated, flattened and abraded on their upper side. Strunian forms have been illustrated (not from the Etroeungt area) by Poty [35] and a systematic monograph is in press (Poty, written communication, 19.07.1995). From the Etroeungt area, Carpentier [36], Vaughan [31, 37], Salée [38] and Dehée [39] gave sketches or short descriptions. Names given below refer to these works but need further systematic revisions. In our collections, the most common species are : *Clisiophyllum omaliusi* HAIME, *Cl.* sp., *Campophyllum flexuosum* (GOLDFUSS), *C.* sp. nov. POTY, 1984 (pl. 2, fig. 4), gen. and sp. nov. A POTY, 1984 (pl. 2, fig. 5), *Paleosmilia aquisgranense* (FRECH, 1885), *Caninia dorlodoti* SALEE *in* Dehée [39] (non Salée, 1912), *Tabulophyllum* sp. nov. POTY, 1984 (pl. 1, fig. 8a-b).

Thus, there is no common species or genus between the Etaoucun section and the Etroeungt section. Ecological conditions could explain this difference: very carbonate-rich and shallow facies at Etaoucun favoured massive stromatoporoids to the detriment of rugose corals, whereas at Etroeungt, more shaly and deeper conditions allowed the development of lamellar stromatoporoids and solitary corals.

Brachiopods (D. Brice)
Brachiopods are absent or extremely rare in the Etaoucun section since we only noticed one athyrid (?) in section (sample CH-ET 55). This absence corresponds to an unfavourable environment in this area, near the D/C boundary, because brachiopods are present and well diversified in other places of the Guilin area, for instance in the Nanbiancun section [40, 41].

In the Le Parcq quarry (Avesnois), the type-locality of the "Etroeungt limestone", brachiopods are present in most of the beds, but they are not common, often disarticulated and not well preserved. Several genera are common to the "Etroeungt limestone" in the Le Parcq quarry and the Nanbiancun section : *Aulacella*, *Schizophoria* (Orthida); *Schuchertella*, *Semiproductus* (Strophoeodonta); *Actinoconchus*, *Cyrtina*, *Cleiothyridina* (Spiriferida) and *Dielasma* (Terebratulida). But the two sections only shared one common species, *Aulacella interlineata* (SOWERBY), a cosmopolitan taxa.

Nevertheless in Avesnois, but further in the north, in the Bocahut quarry, near Avesnes, the environment is more favourable for brachiopods; among the recognized species we must point out the presence of *Semiproductus irregularicostatus* (KRESTOKNIKOV & KARPYCHEV) (det. M. LEGRAND). This guide form is known in the Southern Urals, its type locality, in association with *Siphonodella praesulcata*, in the Kuznetsk, the

Central Kazakhstan and the Tian Shan; it is even reported in South China (Nanbiancun section, Guangxi) where the specimens, which are smaller and with a slightly different ornamentation, could belong to a local variety [41].

The Etroeungt limestone, which corresponds only to the upper part of the Strunian s.l. (table 2), has yielded other characteristic forms which are : *Rugosochonetes* ? sp., *Araratella* sp., *Prospira struniana* (GOSSELET), *P.* nov sp., *Tenticospirifer julii* (DEHEE), *Kitakamithyris* sp., *Composita struniana* (DEHEE), *Lamellosathyris* sp. Moreover, many forms too poorly preserved (spiriferids and strophomenids) are, for the moment, undetermined.

Ostracods (B. Milhau)

The Etaoucun section yielded many ostracods, half of which are unfortunately specifically undeterminable; much more work must be carried out. Most of the species belong to common genera such as *Bairdia*, *Bairdiocypris*, *Shishaella*, *Acratia*, *Microcheilinella*, *Kloedenellitina*, *Paraparchites*, *Bairdiacypris*, *Cavellina and Sulcella*, which is about the same as what was found in the Nanbiancun section [42].

In the upper part of the Dongcun Formation (21 samples, 7 barren, especially in the last 30 meters), the ostracod-fauna is badly preserved, not very abundant, and little diversified. Nevertheless, there are a few exceptions such as CH-ET 3 (30 specimens, 2 species) which contains only *Sinoleperditia* (see Wang [43]) or CH-ET 38 (45 specimens, 15 species) which contains, among other species, *Kloedenellitina triceratina* TSCHIGOVA, 1960, *Kloedenellitina* cf. *binodosa sensu* COEN, 1989, *Marginia* sp. *sensu* COEN, 1989, and also *Sinoleperditia* cf. *dongcunensia* WANG, 1994. All in all, 160 specimens have been picked out and 25 species recognized. Besides Leperditicopida (12 %), the fauna is dominated by Podocopina (44 %), followed by Palaeocopida (24 %) and Metacopina (20 %). This composition, with 36 % of Bairdiacea, 8 % of Paraparchitacea and 16 % of Kloedenellacea, does not correspond to any known type of ecozone; nevertheless the high percentage of Kloedenellacea in association with Leperditicopida indicates a restricted, very shallow, lagoonal-type environment [6, 44].

Table 2 Correlations between Avesnois (France) and Etaoucun (China). 1: Epoch; 2: Stage; 3: Abbreviation in Ardennes, from the geological map of Belgium; 4: Abbreviation *sensu* Conil [13]; 5: Stage *sensu* Conil & Lys [11]; 6: Lithostatigraphy *sensu* Conil [13]; 7: Units in the parastratotype of the "Tranchée d'Avesnelles" [13]; 8: Units at the "Halte de St Hilaire" [13]; 9: Lithostatigraphy in Avesnois [17, 18]; 10 and 11: Foraminifera Zones [14, 20]; 12: Benthic ostracoda Zones [48], *Cryptophyllus* level [14]; 13: Standard conodont Zones; 14: The "Calcaire d'Etroeungt" s.s. [8]; 15: Units in the "Calcaire d'Etroeungt" s.s. [11, 13]; 16: Recognized foraminifera zones in the Etaoucun section; 17: Stratigraphic column of the Etaoucun section; 18: Distribution of stromatoporoids (S) and ostracods (O); O1: Leperditidae association, O2: Second association, O3: Third association.

1	2	3	4	5	6	7	8	9	10	11	12	13	14	15	16	17	18

CARBONIFEROUS

TOURNAISIAN

Tn2

Tn1b — Tn1bβ γ — α

HASTARIAN

Calcaire d'Avesnelles

C f 1 — α" — α' — α

CT01

(sulcata)

Tn1a — Tn1aγ

calcareous — u — t — s — q

Calcaire d'Etroeungt s.l.

Upper DS08

ε'

Yeoyuralng

Cf1e'

hyp.3 hyp.4 — 70 — O3 — S

29-30- 68 25

Df 3ε-Cf1α

Etroeun

55/56

Calcaire d'Etroeungt s.s.

d c b a

DEVONIAN

UPPERMOST FAMENNIAN

Fa2d

Tn1aβ

STRUNIAN

sandy — p — o — n — m — l — k

Schistes de l'Epinette

D f 3

δ

ε

(praesulcata)

Lower DS08

11 8

Df 3γ/δ

hyp.1 — 6 5

Dongoun

4/41 38

34

32

Df 3β

hyp.2

?

O2

O1

UPPER FAM.

Fa2c

Tn1aα

shaly — j — i — h

Schistes de Seins — o — n — m — l — k — j — i

γ — β

(expansa)

DS07

?

?

In the Etaoucun Formation (45 samples, 18 barren, especially in Chinese units 77-78-79), the ostracod-fauna is also badly preserved and generally not very abundant but slightly more diversified. Nevertheless, there are a few exceptions: CH-ET 6 (20 specimens, 14 species) and CH-ET 13 (40 specimens, 17 species) with a lot of Acratidae, Bairdiidae such as *Bairdia* sp. *sensu* COEN, 1989 or *Bairdia gansuensis* SHI & WANG, 1987, and large *Bairdiocypris*, CH-ET 55/15/56 (110 specimens, 30 species) where we must emphasize the recent discovery of Strunian *Sinoleperditia* [6], CH-ET 22/23 (50 specimens, 9 species) with *Evlanella* ? sp. and *Kloedenellitina* sp., CH-ET 24/25 (100 specimens, 23 species) and CH-ET 29 (30 specimens, 9 species) with a lot of Paraparchitidae such as *Shishaella porrecta* (ZANINA, 1956) or *Coelonella* sp. All in all, 440 specimens have been picked out and 90 species recognized; 11 of them, such as *Bairdia* cf. *gibbera* MOREY, 1935 for example, were known in the Dongcun Formation. The fauna is dominated by Podocopina (54 %), followed by Palaeocopida (26 %) and Metacopina (19 %). The ratios of Bairdiacea (46 %), Paraparchitacea (15 %) and Kloedenellacea (7 %) could correspond to the "Bairdiacea and Paraparchitacea ecozone" of Crasquin [44].

In the basal 9 meters of the Yaoyunling Formation, Shangyueshan Member (7 samples), the ostracod-fauna is very well preserved, abundant and highly diversified. For example, CH-ET 31 yielded 170 specimens in which 35 species were identified such as *Pseudoleperditia poolei* SOHN, 1969, *Shishaella opima* ZHANG, 1985, *Bairdia cestriensis* ULRICH, 1891, *Acratia* cf. *acutiangulata* (POSNER, 1960), *Praepilatina* ? sp., and also one species of *Monoceratina* ?. All in all, among 400 picked specimens, more than 60 species have been recognized; about 20 were known in the Etaoucun Formation, 15 of them emerging in samples CH-ET 25 and CH-ET 29 like *Bairdia moreyi* HARRIS & JOBE, 1956, *Paraparchites* sp. or *Shishaella* sp. The fauna is dominated by Podocopina (56 %), followed by Palaeocopida (30 %) and Metacopina (13 %). The ratios of Bairdiacea (53 %), Paraparchitacea (22 %) and Kloedenellacea (7 %) typically correspond to a "Bairdiacea and Paraparchitacea ecozone".

In the Strunian of Avesnois, "tranchée d'Avesnelles" and Le Parcq quarry, Lethiers [45-48] reported 46 species of ostracoda, 18 of which were known in the Upper Famennian ("Schistes de Sains"). Actually, work in progress indicates that the ostracods are more diversified and dominated by Podocopina (52 %); Metacopina and Eridostraca are better represented in the calcareous Strunian (respectively 16 % and 13 %) than in the shaly and sandy Strunian (9 % and 6 %), which is the reverse for Paleocopida (20 % against 33%). The ostracod association also corresponds to the "Bairdiacea and Paraparchitacea ecozone" [44] with 46% of Bairdiacea, 12 % of Paraparchitacea and 11 % of Kloedenellacea, but there is no species in common with the Guilin area fauna.

The comparison of Strunian ostracods from Etaoucun with those found in Avesnois (France) leads to the following three points:

1) the ecozone-type is the same and corresponds to the "Bairdiacea and Paraparchitacea ecozone" [44];
2) most of the genera are found in the two areas but *Cryptophyllus*, common in Avesnois, is absent in Guilin area, and the reverse obtains for *Sinoleperditia*;

3) so far as we know from the systematics, species, even where very similar, are all different.

Conodonts (J.M. Degardin) *and Vertebrate-Microremains* (C. Derycke).
Although many conodonts have been found in the Nanbiancun section [4, 40, 49], parastratotype of the D/C boundary, 15 km north of Etaoucun, nothing has been found in the dozen samples from Etaoucun processed in order to find conodonts and vertebrate microremains. This is probably due to the unfavourable reefal environment, as for brachiopods (see above).

In Avesnois, the Strunian stratotype of the Etroeungt limestone in the Le Parcq quarry has never yielded conodonts (Groessens, pers. comm.). Twenty-three samples from the new cut of the Bocahut quarry have also been processed, but no conodonts or vertebrate microremains have been found.

CONCLUSION

The majority of benthic fossil groups, except foraminifera, present very few taxa common to the two areas; this supports the endemism of the South-China Strunian faunas, already underlined by some authors [21, 50]. South China seems to have been largely isolated from the other land masses during Strunian time. This idea must be taken into account in palaeogeographical reconstructions [29].

Only foraminifera allow a proposal for strong correlation between the Etaoucun section and the Strunian stratotype area (table 2). Foraminifera zones, and their succession, are the same, with the Df3β zone, the Df3γ/δ zone, the Df3ε-Cf1α zone and the Cf1α' zone.

In Avesnois, the Strunian stratotype area, the Upper Famennian/Uppermost Famennian (= Strunian s.l.) boundary is located in the Df3γ zone. In the Etaoucun section, according to our data, this boundary could be placed, as it has been proposed [3], at the top of the Dongcun Formation, where the leperditids almost disappear (hypothesis 1), or could be placed below, in the upper part of the Dongcun Formation, where the stromatoporoids appear (hypothesis 2). If the first hypothesis is confirmed, then reefal levels in samples CH-ET 4 and 41 are of Famennian age.

Following Conil *et al.* [14, 20], the Strunian/Hastarian boundary, in the stratotype area, has to be placed at the base of the Cf1α corresponding to the disppearance of *Quasiendothyra* and the appearance of the typical *Chernyshinella* association. However, there are few precursors of primitive *Chernyshinella* in Strunian, and locally persistance in the lower Hastarian of *Quasiendothyra*. Then the D/C boundary is not so clearly defined when conodonts are absent, and much more work need to be done on foraminifera. Nevertheless it seems clear that the D/C boundary should be located below the unilocular zone Cf1α'. In the Etaoucun section, the D/C boundary has been placed by Conil [4] about 2.5 m below the top of the Etaoucun Formation, just above a rich stromatoporoid-bearing bed (sample CH-ET 29). However *Bisphaera* and *Earlandia* are already present, without *Quasiendothyra*, which is caracteristic of the Cf1α' zone, in sample CH-ET 68 located 1 m below. On the other hand, *Quasiendothyra*, without *Bisphaera* and *Earlandia*, which is caracteristic of the Df3ε-Cf1α, is present in sample CH-ET 30, located above CH-ET 29, that is to say above the postulated D/C boundary. Then, in the Etaoucun section, according to our data, the D/C boundary could be placed where it has been previously proposed, so that stromatoporoids of CH-ET 29 are Strunian (hypothesis 3), or should be lower down, at least below sample CH-ET 68, so that sromatoporoids are Hastarian (hypothesis 4).

Acknowledgment

We would like to thank our Chinese colleagues who organized the two field missions for B. Milhau and B. Mistiaen, especially Dr. YUAN Shiying, president of the Jiaozuo Institute of Technology, and also Dr. YIN Boan from the Geological Survey of Guilin who accompanied us in the field. We also thank Dr. M. LEGRAND (Bordeaux) who determined brachiopod Productoidea and provided to us some references, Dr. E. POTY (Liège) who sent some photos of his new species, and D. HIPPLE and S. BAYART (ISA, Lille) for proof reading.

References

1. Zhou H., Wu Y. and Zhang Z. Devonian paleogeographic framework of Guangxi, South China. *In* McMillan *et al.* (Ed.): Devonian of the World. *Proceedings of the 2nd International Symposium on Devonian System*, 1: 635-643 (1988).
2. Yu C., Bao H., Shen J., Yin B., Zhang S. and Yin D. Devonian reef complexes in Guilin, South China. Second International Congress on Palaeoecology. *Nanjing Institute of Geology and Palaeontology, Academia Sinica. Guidebook for excursion 2*: 67 p. (1991).
3. Yin B., Zhou J., Zhang S., Chen J., Huang J. and Lu H. Etoucun section. The Devonian-Carboniferous Boundary. 11th International Congress of Carboniferous Stratigraphy and Geology. *Guidebook for excursion 6*, Carboniferous Carbonate Sequences in Guangxi, 1, B: 6-7 (1987).
4. Li Z., Guo S., He G. and Yu C. Regional Stratigraphy. *In* YU Chang-min (Ed.): Devonian-Carboniferous Boundary in Nanbiancun, Guilin, China. Aspects and records. *Science Press, Beijing*: 9-18 (1988).
5. B. Mistiaen Découverte du genre *Amphipora* SCHULZ, 1883, dans le Famennien terminal "Strunien" de la carrière du Parcq, à Etroeungt, stratotype du "Calcaire d'Etroeungt" et ailleurs en Avesnois (Nord de la France). *Compte Rendu de l'Académie des Sciences* (in press).
6. B. Milhau Présence de Leperditiidae (Ostracoda) dans le Dévonien supérieur d'Etaoucun (Guangxi, Chine du Sud). Signification paléoécologique. *Geobios* (in press).
7. Hou H. An outline of Famennian Stratigraphy of South China. *Stratigraphy and Paleontology of China*, 1: 49-69 (1991).
8. J. Gosselet Note sur le terrain dévonien de l'Ardenne et du Hainaut. *Bulletin de la Société géologique de France*, 2, 14: 364-374 (1857).
9. J. Gosselet Mémoire sur les terrains primaires de la Belgique, des environs d'Avesnes et du Boulonnais. *Savy éd.*, Paris (1860).
10. E. Mailleux & F. Demanet L'échelle stratigraphique des terrains primaires de la Belgique. *Bulletin de la Société belge de Géologie*, 38 (2): 124-131 (1928).

11. R. Conil & M. Lys Strunien. *In* Cavelier C. and Roger J. (Coord): Comité Français de Stratigraphie. Les étages français et leurs stratotypes. *Mémoire B.R.G.M.*, 109: 26-35 (1980).

12. B. Mamet., G. Mortelmans and P. Sartenaer Réflexions à propos du Calcaire d'Etroeungt. *Bulletin de la Société belge de Géologie*, 74: 41-51 (1965).

13. R. Conil, M. Lys and E. Paproth Localités et coupes types pour l'étude du Tournaisien inférieur. Révisions des limites sous l'aspect micropaléontologique. *Académie royale de Belgique*, Classe Sciences, Mémoires, 4, 2, 15 (4): 1-87 (1964).

14. R. Conil, R. Dreesen, M.A. Lentz, M. Lys and G. Plodowski The Devono-Carboniferous transition in the Franco-Belgian Basin with reference to Foraminifera and Brachiopods. *Annales de la Société géologique de Belgique*, 109: 19-26 (1986).

15. R. Conil & M. Lys Données nouvelles sur les Foraminifères des couches de passage du Famennien au Tournaisien dans l'Avesnois. Colloque de Liège, 1969. *Congrès et colloques de l'Université de Liège*, Stratigraphie du Carbonifère, 55: 241-265 (1970).

16. R. Conil Intérêt de certaines coupes de l'Avesnois dans la séquence classique du Dinantien. *Annales de la Société Géologique du Nord*, 43: 169-175 (1973).

17. R. Conil, H. Pirlet and M. Lys Traits dominants de l'échelle biostratigraphique du Dinantien de la Belgique. *Congrès de Sheffield*, 1967. 1: 45-49 (1969).

18. R. Conil & M. Lys Aperçu sur les associations de Foraminifères Endothyroides du Dinantien de la Belgique. *Annales de la Société Géologique de Belgique*, 90 (4): 395-412 (1967).

19. Wang K. On the Devonian-Carboniferous boundary based on foraminiferal fauna from South China. *Acta Micropaleontologica Sinica*, 4, 2: 161-177 (1987).

20. R. Conil, Th De Putter, Hou H., Wei J. and Wu X. Contribution à l'étude des Foraminifères du Strunien et du Dinantien de la Chine sud-orientale. *Bulletin de la Société belge de Géologie*, 97 (1): 47-61 (1988).

21. L. Hance, Ph Muchez, M. Coen, Fang X., E. Groessens, Hou H., E. Poty, Ph. Steemans, M. Streel, Tan Z., F. Tourneur, M. Van Steenwinkel and Xu S. Biostratigraphy and sequence stratigraphy at the Devonian-Carboniferous transition in Southern China (Hunan Province). Comparison with Southern Belgium. *Annales de la Société Géologique de Belgique*, 116 (2): 359-378 (1993).

22. R. Conil, E. Groessens, M. Laloux, E. Poty and F. Tourneur Carboniferous guide foraminifera, corals and conodonts in the Franco-Belgian and Campine Basins: their potential for widespread correlation. *Courier Forschunginstitut Senckenberg*, 130: 15-30 (1990).

23. D. Vachard & B. Clement L'Hastarien (ex- Tournaisien inférieur et moyen) à Algues et Foraminifères de la zone pélagonienne (Attique, Grèce). *Revue de Micropaléontologie*, 37, 4: 289-319 (1994).

24. B. Mistiaen Identité des genres *Stylostroma* GORSKY 1938 et *Pennastroma* DONG DE-YUAN 1964, Stromatopores du Famennien supérieur (Strunien). *Geobios* (in press).

25. C. W. Stearn Effect of Frasnian -Famennian extinction event on the stromatoporoids. *Geology*, 15: 677-679 (1987).

26. A. E. Cockbain Distribution of Frasnian and Famennian Stromatoporoids. *Memoir Association Australasian Palaeontologists*, 8: 339-345 (1989).

27. C. W. Stearn, M. K. Halim-Dihardja and D. K. Nishida An Oil-Producing Stromatoporoid Patch Reef in the Famennian (Devonian) Wabamun Formation, Normandville Field, Alberta. *Palaios*, 2: 560-570 (1987).

28. D. Le Maitre Description des Stromatoporoides de l'assise d'Etroeungt. *Mémoire de la Société géologique de France*, 9, 1 (20): 1-32 (1933).

29. B. Mistiaen, B. Milhau, D. Brice, J. M. Degardin, C. Derycke, Hou H., J. C. Rohart, D. Vachard and Wu X. Uppermost Famennian fauna from Etroeungt (Avesnois, North of France) and Etaoucun (Guangxi, South China). Palaeogeographical implications. *Annales de la Société Géologique du Nord* (in press).

30. Yu C. Lower Carboniferous corals of China. *Palaeontologica Sinica*, B, **12** (3): 211 p. (1933).

31. A. Vaughan Correlation of Dinantian and Avonian. *Quarterly Journal of the Geological Society of London*, **71**: 1-52 (1915).

32. Hou H. Early Carboniferous brachiopods from the Mengkungao Formation of Gieling, Central Hunan and discussion of the Lower boundary of the Carboniferous. *Professional Papers, Acad. Geol. Sci.*, Ser.B, (1): 116-146 (1965).

33. Lin B. & Xu S. Late Devonian to Tournaisian rugose corals from South China and palaeontological events. *Courier Forschungsinstitut Senckenberg*, **172**: 23-33 (1994).

34. Yu C. Corals. *In* YU Chang-min (Ed.): Devonian-Carboniferous Boundary in Nanbiancun, Guilin, China. Aspects and records. *Beijing Science Press*: 165-195 (1988).

35. E. Poty Rugose corals at the Devonian-Carboniferous boundary. *Courier Forschungsinstitut Senckenberg*, **67**: 29-35 (1984).

36. A. Carpentier Contribution à l'étude du Carbonifère du Nord de la France. *Mémoires de la Société géologique du Nord*, **7**, 18: 434 p. (1913).

37. A. Vaughan Note on *Clisiophyllum ingletonense* sp. nov.. *Proceedings of the Yorkshire geological Society* , **17**: 251-255 (1912).

38. A. Salee Contribution à l'étude des polypiers du calcaire carbonifère de la Belgique.- II. Le groupe des Clisiophyllides. *Mémoires de l'Institut géologique de l'Université de Louvain*, **2**: 179-293 (1913).

39. R. Dehee Description de la faune d'Etroeungt. Faune de passage du Dévonien au Carbonifère. *Mémoire de la Société géologique de France*, n. série, **5**, (2), 1-65 (1929).

40. Li Z., Lu H. and Yu C. Description of the Devonian-Carboniferous Boundary Sections. *In* YU CHANG MIN (Ed.): Devonian-Carboniferous Boundary in Nanbiancun, Guilin, China. Aspects and records. *Beijing Science Press*: 19-36 (1988).

41. Xu H. & Yao Z. Brachiopoda. *In* YU Chang-min (Ed.): Devonian-Carboniferous Boundary in Nanbiancun, Guilin, China. Aspects and records. *Beijing Science Press*: 263-326 (1988).

42. Wang S. & G. Becker Paleoecology of late Devonian and early Carboniferous Ostracodes from Guilin and its vicinity, Guangxi. *In* JIN Yu-gan, WANG Jun-geng et XU Shan-hong (Ed.): Palaeoecology of China. *Nanjing University Press*, **1**: 66-86 (1991).

43. Wang S. A new Leperditiid Tribe Sinoleperditiini (Ostracoda) from Devonian of South China. *Acta Paleontologica Sinica*, **33**, 6: 698-719 (1994).

44. S. Crasquin L'écozone à Bairdiacea et Paraparchitacea (Ostracoda) au Dinantien. *Geobios*, **17** (3): 341-348 (1984).

45. F. Lethiers Ostracodes de la limite Dévonien-Carbonifère dans l'Avesnois. *Compte Rendu de l'Académie des Sciences de Paris*, D, **278**: 1015-1017 (1974).

46. F. Lethiers. Nouveaux ostracodes du passage Dévonien-Carbonifère de la région type. *Compte Rendu de l'Académie des Sciences de Paris*, D, **279**: 1613-1617 (1974).

47. F. Lethiers. Révision de l'espèce *Bairdia (Orthobairdia ?) hypsela* Rome, 1971 (Ostracoda) du Strunien ardennais. *Annales de la Société Géologique du Nord*, **45**: 71-77 (1975).

48. F. Lethiers. Zonation du Dévonien supérieur par les Ostracodes (Ardenne et Boulonnais). *Revue de Micropaléontologie*, **27** (1): 30-42 (1984).

49. Yu C. Attaining a common language of stratigraphy. Basic evaluation of Nanbiancun Section. *In* YU Chang-min (Ed.): Devonian-Carboniferous Boundary in Nanbiancun, Guilin, China. Aspects and records. *Beijing Science Press*: 330-336 (1988).

50. F. Lethiers. Paléogéographie des faunes d'Ostracodes au Dévonien supérieur. *Lethaia*, **16**: 39-49 (1983).

Proc. 30ᵗʰ Int'l. Geol. Congr., Vol. 12, pp. 95-107
Jin and Dineley (Eds)
© VSP 1997

Palynological Study of the Devonian - Carboniferous Boundary in the Vicinity of the International Auxiliary Stratotype Section, Guilin, China.

YANG, WEIPING[1], ROGER NEVES[2]

[1] *Nanjing Institute of Geology & Palaeontology, Academia Sinica, Nanjing 210008. P.R.China*
[2] *Centre for palynological studies, Sheffield University, Sheffield S1 3JD, UK.*

Abstract

Palynological data from Guping Section in the vicinity of the International Auxiliary Stratotype Section of the Devonian - Carboniferous Boundary, Guilin, China, have been obtained for the first time. Two miospore zones have been established: Pmr, Pml. These two assemblages, which contain latest Devonian palynomorphs, can be correlated very well with those in Byelorussia, Poland, and West Europe, as well as Central Hunan in China. Consequently, four miospore evolutionary stages in the above mentioned areas are proposed, i.e. (1) *Retispora lepidophyta* dominant stage; (2) *R. lepidophyta* and *Vallatisporites pusillites* coexisting stage; (3) *V. pusillites* dominant stage; (4) *R. lepidophyta* reoccurring stage. The Pmr and Pml zones in Guilin may be attributed to stage (3) and stage (4) respectively. The disadvantage of using the disappearance of *R. lepidophyta* and *V. pusillites* as the criterion for recognizing Devonian - Carboniferous boundary has also been discussed.

Keywords: miospores, Guilin, China, latest Devonian

INTRODUCTION

Since 1988 at the Courtmacsherry Meeting, the Nanbiancun section in Guilin of Guangxi province has been selected as the International Auxiliary Stratotype Section due to its extraordinary petrographical facies of carbonate platform marginal slope with abundant fossils. Conodonts, brachiopods, trilobites and ostracods have been exhaustively studied in the past but until this study was carried out, there were no palynological data available. A 4mm thick black shale (49th layer) just below the Devonian - Carboniferous boundary has been considered as the equivalent horizon of Hangenberg shale in West Germany and Changshun shale in Guizhou, China [35]. Many attempts to recover palynomorphs from this black shale have been carried out, but all failed. However, a section called Guping 10km away from Guilin town contains a rock suite of the depression facies. Very abundant and well-preserved latest Devonian miospores were first extracted from the dark grey muddy and silty shales. This discovery of palynomorphs in Guilin fills a gap in our palynological knowledge and enriches the biostratigraphical study in the Guilin area.

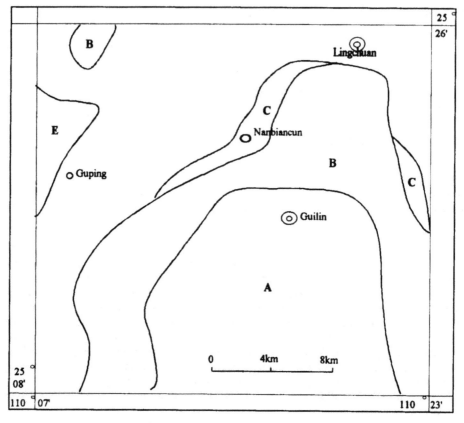

Figure 1. Location and division of late Devonian sedimentary facies in Guilin and its vicinity (after Yu *et al*, 1988)
A. Restricted platform facies; B. Open platform to marginal platform facies; C. Front Slope of platform facies;
D. Depressional facies; E. Depressional terrigenous bank facies.

PALYNOSTRATIGRAPHY

Miospore assemblages in uppermost Devonian of Guping section, Guilin
Fifteen palynological samples were collected from the 1st member of the Luzhai
Formation, which overlays the silicates of the Liujiang Formation. The lithology of the
Luzhai Formation consists of gray and dark gray interbedded sandstone, mudstone, and
shale. The samples were processed using conventional palynological techniques, eight
samples were productive and contain well-preserved palynomorphs. The palynological
assemblages dated as latest Devonian in age are introduced as follows:

The palynomorphs extracted from samples GP-2 to GP-8 are very significant and
therefore named provisionally the *Vallatisporites pusillites, Tumulispora malevkensis*
and *Vallatisporites robustospinosus* Pmr assemblage. Their combined occurrence
indicates a latest Devonian age. The lowest sample, i.e. sample GP-2, the
Lophozonotriletes complex and the index fossils *Verrucosisporite s nitidus,* and

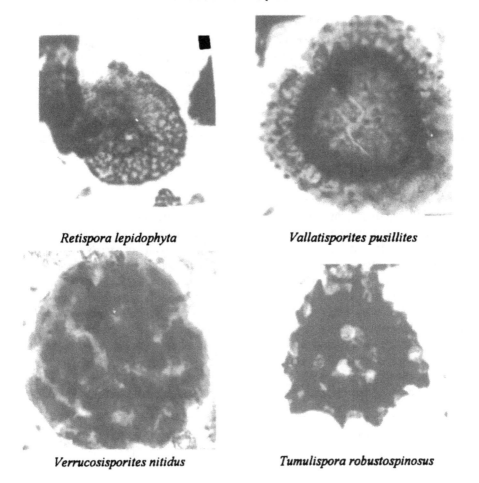

Retispora lepidophyta

Vallatisporites pusillites

Verrucosisporites nitidus

Tumulispora robustospinosus

Figure 2. Some key species (× 800) of latest Devonian miospore from Guping section in Guilin, China

Tumulispora malevkensis are predominant. This assemblage also yields abundant cingulate forms such as *Densosporites* sp-A, *D.* sp-B, and *Anulatisporites* as well as species of the *Lophozonotriletes* complex which seem to be cingulate, but are in fact non-cingulate and acamerate. The following species are also associated within this assemblage. They are *Rugospora* sp. *Microreticulatisporites* sp *Punctatisporites* sp., *Vallatisporites* sp., *Camptotriletes* cf. *prionatus*, *Knoxisporites* cf. *literatus*, *Tumulispora monstruosa*, *Gorgonispora* sp., *Dictyotriletes* sp., *Calamospora* sp., *Cristatisporites* sp., and *Tholisporites mirabilis*. Typical Pmr elements (*Vallatisporites pusillites*, *Tumulispora malevkensis* and *Vallatisporites robustospinosus*) occur in samples GP-3 to GP-8. They also contain: *Tumulispora monstruosa*, *T. variverrucata*, *Apiculiretusisporites hunanensis*, *Grandispora* sp-A, *G.* cf. *cornuta*, *Acanthotriletes* sp., *Densosporites spitsbergensis*, *D.* sp., *Calamospora* sp., *Plicatispora* sp., *Punctatisporites planus*, *P. irrasus*, *P.* sp., *Apiculatisporis heteroconus*, *A.* sp., *Apiculiretusispora fructicosa*, *A. rarispinosa*, *Corystisporites* sp., *Petrotriletes* sp.,

Convolutispora permixta, C. major, Pustulatisporites cf. *dolbii, Cymbosporites* cf. *magnificus, Auroraspora macra, A.* cf. *corporiga, Planisporites* sp., *Rugospora* cf. *corporata, R.* sp., *Camptotriletes paprothii, Latosporites* sp., *Microreticulatisporites* sp., *Bascaudaspora* sp., *Vallatisporites microspinosus, Knoxisporites triangularis, K.* sp., *Discernisporites micromanifestus, Spelaeotriletes microspinosus, Retusotriletes communis, Radiizonatus* sp., *Kraeuselisporites* cf. *hibernicus, Crassispora catena.*

Sample GP-9 may be distinguished from the others in that some elements (partially broken) of the latest Devonian marker *Retispora lepidophyta* are also preserved. Therefore, the assemblage is named as Pml zone due to the occurrence of *Retispora lepidophyta* var.*minor* in percentages of 2-3%.

Correlation with Marine Fauna

Some of the marine fauna such as ammonoids, ostracods, and conodonts were studied before the current miospore study in Guping section was made. The original 60th bed of Luzhai formation in Guping section yielded abundant ostracods, *R.(R).costata, R.(R.)striatula, Maternella hemisphaerica,*(Regional Geological Survey Report of Guilin Urban, 1:50000,1988) which were the representative elements of the *hemisphaerica - dichotoma* biozone at the top of Famennian in the Rheinisches district of West Germany, and also occurred in Lufu of Ludan in West Guangxi province. The ammonoid *Tornoceras* ? sp. was previously only reported elsewhere in the middle and upper Devonian. The faunal horizons stratigraphically correspond to the upper part of miospore Pmr zone and Pml zone *Siphonodella dulplicata*, the earliest Carboniferous conodont biozone occurred just above the Pml miospore zone in the Guping section [35]. The detailed correlation between miospores and fauna are illustrated in table-1.

	Conodonts	Ostracoda	Ammonoids	Spores
Carboniferous	*Siphonodella dulplicata*	*Maternella seilerensis* *M.circumcostata* *Ungerella* sp.		
Devonian		*Richterina (Richterina) costata* *Maternella* *hemisphaerica*	*Tornoceras* sp.	Pml: *Vallatisporites pusillites* *Tumulispora malevkensis* *Retispora lepidophyta* var.*minor* Pmr: *Vallatisporites pusillites* *Tumulispora malevkensis* & *T. robustospinosus*

Table 1. Correlation between latest Devonian miospores and fauna in Guping, Guilin

Table 2. A tentative correlation chart of Devonian - Carboniferous miospore assemblages in China

Region	Author	Formation(s)	Carboniferous zones (Tn2b-c / Tn2a / Tn1b)	Devonian zones (Tn1a-b)
Guilin	This paper	1st member Luzhai Formation	Pnl, Pnr	
W. Yunnan	Yang, W. 1993	Longba Formation	PC, BP	LE
N. Tarim	Gao, L. 1991	Sand well No. 10	PC, HD, VI	
Jiangxi	Wen et al 1993	Farxia Fm., Huanggang Fm., Liujiateng Fm.	TT	MX
W. Zhejiang	He et al 1993	Hsitun Formation	DP	LC, LH
Lower Yangzi	Gao, L. 1991	Leigutai Member	BP, VI	PL, LL
Lower Yangzi	Ouyang 1989	Jinling Fm., Leigutai Member	DP, MD	LC, LH
Jurong Jiangsu	Ouyang et al 1987	Jinling Fm., Leigutai Member	DC	LC, LH
Baoying Jiangsu	Ouyang et al 1987	Jinling Fm., Leigutai Member	DM	LH
S.E. Guizhou	Gao, L. 1991	Tangbagou Fm., upper Gelaohe Fm, lower Gelaohe Fm.	TM, VI*	PN
Central W.Hubei N.W.Hunan	Gao, L. 1992	Changyang Formation, Tizikou Formation	PC, VI*	LN, LE
Central Hunan	Gao, L. 1990	Menggongao Formation, Shaodong Formation	FM, VI*	LN, LE, LL
Central Hunan	Yang, Y 1987	Malanbian Formation, Menggongao Formation	NV	PL
Tibet	Gao, L 1988	Yali Formation, Zhangdong Formation	BM, VI*	LN, PL
West Europe	Higgs et al 1988		PC, BP, HD, VI	LN, LE, LL
Stratigraphy			Tn2b-c, Tn2a, Tn1b	Tn1a-b

PREVIOUS WORK

South China

Devonian - Carboniferous boundary miospore studies in China started in the late 1970s. The *Retispora lepidophyta* miospore assemblage was first reported in horizons associated with the fish *Sinolepis* of the Leigutai Formation (upper Wutong Group) in Nanjing, Jiangsu, as well as with the fish *Bothriolepis* of the Xikuangshan Formation at the top of upper Devonian in Hunan [24]. Since then, the miospore study of Devonian - Carboniferous in China has entered a new phase. Devonian - Carboniferous transition miospore assemblages have been subsequently reported in the Zhangdong Fm. of Qomolangma Mt. and Nielamu in Tibet (Gao, 1983) [5]; Menggongao Fm. of the Xikuangshan area in Hunan (Hou, 1982) [17]; Getongguan bed and Dawoba Fm. of Muhua in Guizhou (Gao, 1985) [16]; Zhewang Fm. and Gelaohe Fm. of Southeast Guizhou (Gao, 1991) [9]; Tizikou Fm. and Changyang Fm. of Northwest Hunan (Gao, 1992) [10]; Shaodong Fm. and Menggongao Fm. of central Hunan (Gao, 1990) [6]; Sand No.10 well from drilling data of north Tarim basin (Gao, 1991) [7]; Wutong Group of Jurong in Jiangsu (Ouyang *et al.*, 1987) [20]; Baoying district in Jiangsu (Ouyang, 1987) [21]; lower Yangtze river of Jiangsu (Ouyang *et al.*, 1989) [22]; Wutong, Laokan, and Cishan formations of lower Yangtze river (Gao, 1991) [8]; Hsihu Fm. of Fuyang in West Zhejiang (He *et al.*, 1993) [12]; Fanxia Fm. and Huangtang Fm. of Quannan Xiaomu in Jiangxi (Wen *et al.*, 1993) [31]; Longba Fm. of Sipaishan Mt. in Gengma, West Yunnan (Yang *et al*, 1991; Yang, 1993) [32,33]. The miospore studies of Devonian - Carboniferous transitions in the above areas made a great contribution to resolving the age of relevant strata which had been uncertain for a long time. For example, in the Wutong Group of lower Yangzte River Valley, as well as in the Menggongao Formation of Hunan, some earliest Carboniferous miospores were found. A tentative correlation chart of Devonian - Carboniferous miospore assemblages from some of the above-mentioned areas in South China is illustrated in Table-2.

Latest Devonian Miospores Correlation

The disappearance of *R. lepidophyta* and *V. pusillites* to define the palynological Devonian - Carboniferous boundary has been widely used in practice [11,25,28]. *V. pusillites* was first described by Kedo in 1957 [18] as *Hymenozonotrileges pusillites*, from the Malevka Formation of the Pripyat basin in Byelorussia and where it was associated with *R. lepidophyta*. In 1970, Dolby and Neves [3] modified this original name to *V. pusillites*. Streel (1970) [29] and Mcgregor (1970) [19] discussed the vertical/stratigraphical and geographical distribution of these two species. They found tremendous similarities in both stratigraphical and geographical distribution of these two species, ranging from Fa2d to the lower part of Tn1b. *V. pusillites* and its resembling elements have widely been reported from Upper Devonian in South China, e.g. in Nielamu of Tibet(Gao, 1983) [5]; from Getongguan Formation in Muhua of Guizhou (Hou *et al*, 1985) [16]; in Central Hunan (Yang.Y, 1987) [34], from the lower and middle part of Leigutai member of Wutong Fm. in Jiangsu (Gao, 1990 [8]; Ouyang and Chen, 1987 [22]); from Northwest margin of Zungur basin of Xinjiang (Zhou, 1989, M.Sc. thesis). *R. lepidophyta* was also erected by Kedo in 1957 [18]. He used the name *Hymenozonotrileges lepidophytus*, but later, Playford (1976) [27] formally proposed the combination of *R. lepidophyta*. In 1967, Owens and Streel [23] had already noticed the remarkable distribution of *R. lepidophyta*. Up to now, more than forty localities in eighteen countries belonging to five continents have reported the occurrence of *R. lepidophyta*, which ranged from the middle and late Famennian (Fa2d) to the early Tournaisian (Tn1a and Tn1b). Although *R. lepidophyta* does occur in early Carboniferous strata, these broken rare specimens have proven to be the result of reworking (Playford, 1976 [27], Yang, 1993 [33]).

The two assemblages from Guilin; Pml: *R. lepidophyta* var. *minor - V. pusillites - Tumulispora malevkensis* and Pmr: *V. pusillites - T. malevkensis & T. robustospinosus*, can be correlated with the palynological assemblages from Muhua and Dushan of Guizhou as well as those in central Hunan in South China, from the Pripyat Depression area of Byelorussia, from Western Europe and Poland. Yang Yunchen (1987) [34] established two miospore zones when he described the Malanbian section from Xinshao in Hunan: *Vallatisporites pusillites - Retispora lepidophyta* (PL) Zone and *Verrucosisporites nitidus - Vallatisporites vallatus* (NV) Zone. Furthermore, the PL zone has been subdivided into three subzones. The lower subzone is dominated by sole occurrence of *R. lepidophyta*. The middle subzone is characterised by the concurrence of *R. lepidophyta* and *V. pusillites*. *V. pusillites* is the prevailing element in the upper subzone with the absence of *R.lepidophyta*, fossil present at this horizon is the coral: *Cystophrentis* sp. The vertical changes in abundance of *R. lepidophyta* from bottom to top are abundance (lower) to decrease (middle) and near absence (upper). Similar changes in miospore assemblages occur in the uppermost Devonian of the Pripyat Depression in the former U.S.S.R [2,18]. Every species of *Vallatisporites* developed rapidly in the Velizh bed of kalinovsky sequence. However, at the top of this sequence *V. pusillites* , *V. vallatus* and some relevant elements become rare. Even tiny and slender specimen of *R. lepidophyta* with low quantity decreased quickly. However, some species of *Tumulispora* increased abruptly. Therefore, this assemblage was called *Vallatisporites pusillites - Tumulispora malevkensis* (PM), which is the top miospore zone succeeding *Retispora lepidophyta - Knoxisporites literatus* (LL), *Retispora lepidophyta - Hymenozonotriletes explanatus* (LE), and *Vallatisporites pusillites - R. lepidophyta - H. explanatus* (PLE) zones of late Devonian in the Pripyat Depression. Conodonts *Siphonodella praesulcata, Pseudopolygnatus fusiformis* coexisted in this assemblage. In 1992, Higgs [15] had already noticed the differentiation of LN (*Retispora lepidophyta - Verrucosisporites nitidus*) zone in the lower and upper parts when he discussed the palynological assemblages in the new Stockum trench II and the Hasselbachtal borehole. The lower LN zone was of typical LN character in terms of Irish miospores (Standard Western Europe) [14]. The upper LN zone, however, was a transitional LN in which *R. lepidophyta* was quite rare (less than 1%). Such a low percentage of *R. lepidophyta* would easily lead to the conclusion of absence or even disappearance if careless. Such a judgement would influence the proper division of Devonian - Carboniferous boundary. In brief, the latest Devonian miospores in Guilin could be correlated with the PM subzone (*V. pusillites - Tumulispora malevkensis*) of P zone (*V. pusillites*) at the top of upper Devonian in central and east of Russia platform. *Verrucosisporites nitidus* was first erected by Naumova as *Lophotriletes grumosus*, later Playford 1963 [26] modified it to be *V. nitidus*, which ranged stratigraphically from the

Table 3. The correlation chart of latest Devonian miospores between Guilin, Central Hunan and elsewhere

STRATIGRAPHY		WESTERN EUROPE		ZONAL SUBDIVISIONS		C & E of Russian Platform		Central Hunan		Guilin, S.China		
		ZONES	LITHOLOGY	Ammonoids	Conodonts					zones	Sample No.	Lithology
DEVONIAN	FAMENNIAN											
		Tn1b										
CARBONIFEROUS	HASTARIAN	Tn2a										
		LV	Epinette	WOCKLUMERIA	Bispathodus costatus	R.lepidophyta (L)	Lty	V.pusillites- R.lepidophyta (PL)				
		LL	Etroeungt	WOCKLUM KALK			Ltn		L.subzone		GP-2	
		LE		Cy.evoluta			PLE		M.subzone		GP-3	
		LN	HANGENBERG SCHIEFER	Imitoceras	Protognathodus	V.pusillites (P)	PM	U.subzone	Pmr	GP-8	
		VI	Hastiere	Acutimitoceras	S.sulcata	Pmi		R.lepidophyta	Pml	GP-9	
			HANGENBERG KALK	GATTENDORFIA	S.duplicata	T.malevkensis (M)		V.vallatus			

1st Member Luzhai Formation

LN zone to CM zone in Western Europe [13,14], i.e. from early Tournaisian (bottom of Tn1b) to top of Tournaisian (Tn3). *V. nitidus* is the index fossils of miospore zone LN (*R. lepidophyta - V. nitidus*) at the end of latest Devonian in Western Europe. In the Russian platform, *V. nitidus* was also found in P zone, therefore, it had been correlated with LN zone in W.Europe [1]. The original definition of LN zone was based upon the first occurrence of *V. nitidus* and the gradual disappearance of *R. lepidophyta*. The miospores in Guping section (Guilin) reflect exactly this kind of miospore evolutionary trend. In the Pmr zone, miospores include abundant *V. pusillites - Tumulispora malevkensis & T.robustospinosus, Verrucosisporites nitidus. R. lepidophyta* var. *minor* only occurred at the end of the Upper Devonian with a low percentage of about 2-3%. A similar phenomenon occurs at the Berchogur section in the Mugodzhar Mountains of the former USSR. *R. lepidophyta*, mainly a few of *R. lepidophyta* var. *tenera*, is present in the latest Devonian ML zone (*Tumulispora malevkensis - Retispora lepidophyta*). While in Byelorussia *R. lepidophyta* var. *minor* and *T. malevkensis* occurred in the PM zone [2]. The interesting thing is that Fang Xiaosi 1993 [4] claimed finding *R. lepidophyta* at the top of PL zone in central Hunan. This species was not recorded from this horizon in Yang Yunchen's (1987) [34] paper. He also proposed a new formation-Tianxin Formation due to lithological differences. If his claim is true, the courses of miospore evolution between Guilin and central Hunan are quite comparable.

PALYNOSTRATIGRAPHY OF THE DEVONIAN-CARBONIFEROUS BOUNDARY

Four Miospore Evolutionary Stages in The Latest Devonian (Figure 3)
According to the above analysis and correlation, there are some consistent trends in miospore evolution in the latest Devonian of Guilin, Central Hunan, and Guizhou in China; Byelorussia; Poland and western Europe. Four miospore evolutionary stages have been established. The first stage is characterised by the predominance of *Retispora lepidophyta* which reaches up to 14 %, e.g. in the lower part of PL zone in central Hunan; in the L zone of eastern and central Russia platform [1]; in the Ra zone of western Pomeranian in Poland [30. The second stage is characterised by the coexistence of *Retispora lepidophyta* and V*allatisporites pusillites*, e.g. in the middle part of the PL zone in central Hunan; in the PLE subzone of the P zone in the Russian platform, and in the LL zone and LE zone of western Europe. *Vallatisporites* is dominant prevails in the third stage, e.g. in the upper part of the PL zone in central Hunan; in the PM subzone of the P zone in the Russian platform. The Pmr assemblage in Guilin should be assigned to this stage and the most of the LN zone in western Europe belongs to this stage. However, Poland apparently lacks of this stage. The fourth stage is characterised by the reoccurrence of *Retispora lepidophyta*, generally as two subspecies of *R. lepidophyta*, i.e. *R. lepidophyta* var.*tenea* in the ML zone of the 2b member in the Berchogur section of Mugodzhar Mt., former USSR [1]; *R. lepidophyta* var.*minor* in the Pml zone in Guilin. Thus, the Pmr and Pml assemblages in Guilin could be attributed to stage 3 and stage 4 respectively, reflecting the latest palynomorphs evolution in late Devonian. (Figure 3)

Comments on The Devonian - Carboniferous Palynological Boundary
The following focuses on two issues related to the Devonian - Carboniferous

palynological boundary; *the confusion* between the absence and disappearance of index miospores in the latest Devonian and *the innate shortage* of index species at the Devonian - Carboniferous palynological boundary.

		Key Species		Guilin	Central Hunan	Central & Eastern Russia	Western Europe
Carboniferous	Tn2a	*Umbonatisporites distinctus*				platform	VI
	Tn1b				*V.vallatus*	Pmi	
Devonian			*R. lepidophyta* reoccurring	Pml assemblage	*R.lepidophyta*	PM	LN
			V.pusillites dominant	Pmr assemblage	U.subzone		
			R.lepidophyta V.pusillites coexisting		M.subzone	PLE	
			R.lepidophyta dominant		L.subzone	Ltn	LE

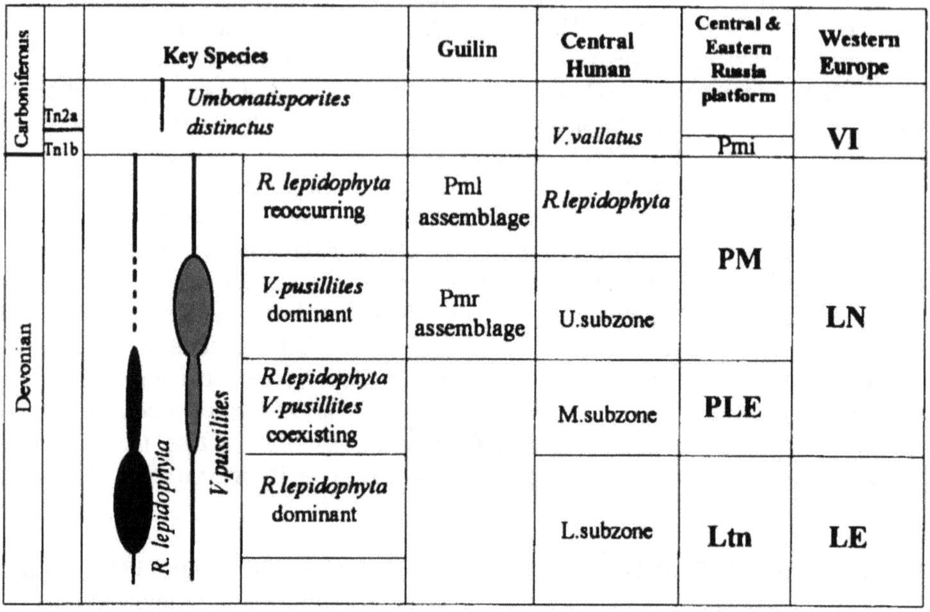

Figure 3. Four evolutionary stages of miospores within the latest Devonian

The Confusion

In practice, both in China and abroad, the criteria of *R. lepidophyta* and *V. pusillites* as the Devonian - Carboniferous palynological boundary usually creates some confusion because of the difficulty in defining the contact between the very rare occurrence of *R. lepidophyta* and its actual absence. In other words, one cannot distinguish the disappearance or absence of *R. lepidophyta* when *R. lepidophyta* is less than 1% such as in the new Stockum trench II and the Hasselbachtal borehole (Higgs *et al*, 1992) [15]. One more example is the Malanbian Section from Xinshao county in central Hunan. Gao Lianda (1990) [6] dated palynologically the Menggongao Formation as early Carboniferous due to the lack of *R. lepidophyta* at the top of the Upper Devonian. In fact, Yang Yunchen (1987) [34], and later Fang Xiaosi (1993) [4], found rare specimens of *R. lepidophyta* throughout the Menggongao Formation.

The Innate Shortage

Based upon both the modern punctuated equilibrium theory about speciation and stratigraphical division practice in many periods, most boundaries between time-rock units rely on the first appearance of some new species, including fauna and floral. Therefore, the palynological boundary between the Devonian and the Carboniferous based upon the disappearance of *R. lepidophyta* and *V. pusillites* is seemingly of innate scarcity.

At the level of the conodont boundary between the Devonian and the Carboniferous, there seemingly appear no new species of miospores till Tn2a in which *Umbonatisporites (Dibolisporites) distinctus* occurs widely for the first time in Western Europe, Australia, and China.

CONCLUSION

A palynological investigation has been carried out recently in the vicinity of the International Auxiliary Stratotype Section of the Devonian - Carboniferous Boundary, Guilin in China. Two latest Devonian miospore assemblages were established for the first time in this area, which are quite comparable to the time-equivalent assemblages from central Hunan, Guizhou, the lower Yangzte River area in South China, and Byelorussia, Poland, and Northern 'Rheinisches Schiefergebirge', Germany. Four miospore evolutionary stages in the above mentioned areas are proposed, i.e. (1) *Retispora lepidophyta* dominant stage; (2) *R.lepidophyta* and *Vallatisporites pusillites* coexisting stage; (3) *V.pusillites* dominant stage; (4) *R.lepidophyta* reoccurring stage. The Pmr and Pml zones in Guilin have been attributed to stage 3 and stage 4 respectively. Some observations can be made on the Devonian - Carboniferous palynological boundary concerning the confusion between the absence and disappearance of index fossils at the latest Devonian where *R. lepidophyta* occurs in extremely low percentages as well as the innate shortage of the Devonian - Carboniferous palynological boundary based upon the disappearance of *R. lepidophyta* and *V. pusillites*.

Acknowledgements

We thank John Utting for his critical reviews which helped to substantially improve the clarity and organization of the manuscript.

REFERENCES

1. V.I. Avchimovitch. Miospore systematics and stratigraphic correlation of Devonian - Carboniferous boundary deposits in the European part of the USSR and Western Europe. *Cour.Forsch.-Inst.Senckenberg* 100 169-191 (1988).
2. V.I. Avchimovitch. Zonation and spore complex of the Devonian and Carboniferous boundary deposits of the Pripyat depression (Byelorussia). *Annales de la Société géologique de Belgique.* .115 (2), 425-451 (1992).
3. G. Dolby and R. Neves. Palynological evidence concerning the Devonian - Carboniferous boundary in the Mendips, England. *C.R. 6th. Congr. Avanc. Etud. Stratigr. Géol. Carb.* (Sheffield,1967).2, 631-646 (1970).
4. X. Fang and M. Streel. New development in the division of Devonian - Carboniferous boundary in central Hunan.*Chinese Science Bulletin* 38.(8), 732-736 (1993).
5. L. Gao. The finding of latest Devonian miospores at Nielamu in Tibet and its geological significance. *Contribution to the Geology of the Qinghai- Xizang (Tibet) plateau* 8, 183-218 (1983).

6. L. Gao. Miospore zones in the Devonian - Carboniferous boundary beds in Hunan and their stratigraphical significance.*Geological Review* **36** (1), 58-68 (1990).

7. L. Gao. The discovery of spores of early Carboniferous age in the Talimu Basin and its geological significance.*Chinese Geology* **12**, 27-28 (1991).

8. L. Gao. New development in the study of late Devonian to the early Carboniferous sporopollen stratigraphy on the lower reaches of Changjiang River. *Chinese Geology* **8**, 28-30 (1991).

9. L. Gao. Latest Devonian - Early Carboniferous miospores and Devonian - Carboniferous boundary in Southeast Guizhou.*Guizhou Geology* **8**. (1), 59-71 (1991).

10. L. Gao. Palynostratigraphy of Devonian - Carboniferous transitional beds in W.Hubei and Northwest Hunan. *Chinese Academy of Geology. Institute of Geology. Memoirs* **23**, 171-192 (1992).

11. T.N. George. A correlation of Dinantian rocks in the British Isles. *Geol. Soc. Lond. Spec. Rep.* **7**, 1-87 (1976).

12. S. He and S. Ouyang. Spore assemblages from Devonian - Carboniferous transitional beds of Hsihu Formation,Fuyang, W.Zhejiang.*Acta Palaeontologica Sinica* **32**. (1), 31-48 (1993).

13. K. Higgs and M. Streel. Spore stratigraphy at the Devonian - Carboniferous boundary in the Northern 'Rheinisches Schiefergebirge' Germany . *Cour. Forsch.-Inst. Senckenberg* **67**, 157-180 (1984).

14. K. Higgs, G. Clayton and J.B. Keegan. Stratigraphical and Systematic Palynology of the Tournaisian Rocks of Ireland. . The Geological Survey of Ireland. *Special Paper Number 7,* (1988).

15. K. Higgs, M. Streel, D. Korn and E. Paproth. Palynological data from the Devonian - Carboniferous boundary beds in the new Stockum trench II and Hasselbachtal borehole. Northern Rhenish Massif, Germany.*Ann. Soc. géol. Belg.* 115.(2), 551-557 (1992).

16. H. Hou. Muhua sections of Devonian - Carboniferous boundary beds. Geological Publishing House, (1985).

17. J. Hou. Miospore assemblages of Devonian - Carboniferous transitional beds in Xikuangshan district, Central Hunan. *Chinese Academy of Geology, Institute of Geology. Memoirs .9*, 81-92 (1982).

18. G.I. Kedo. Spores from the supra-salt Devonian deposits of the Pripyat depression and their stratigraphical significance.*Rep. Palaeont. Strat. Byelorussia. S.S.E.*4, 3-121 (1957).

19. D.C. McGregor. *Hymenozonotriletes lepidophytus* Kedo and associated spores from the Devonian of Canada. *In "Colloques sur la Stratigraphie du Carbonifère" Congrès et Colloques Univ. Liège* **55**, 315-326 (1970).

20. S. Ouyang and Y. Chen. Miospore assemblages from the Devonian - Carboniferous transition in Jurong of Southern Jiangsu with special reference to the geological age of the Wutung Group. *Memoirs of Nanjing Institute of Geology & Palaeontology, Academia Sinica* **23**, 1-120 (1987).

21. S. Ouyang and Y. Chen. Miospores of the Famennian and Tournaisian deposits from a borehole in the Baoying district, central Jiangsu. *Acta Micropalaeontologica Sinica* **4**.(2), 195-216 (1987).

22. S. Ouyang and Y. Chen. Palynology of Devonian - Carboniferous Transition sequences of Jiangsu, E.China.*Palaeontogia Cathayana* **4**, 439-473 (1989).

23. B. Owens and M. Streel. *Hymenozonotriletes lepidophytus* Kedo, its distribution and significance in relation to the Devonian-Carboniferous boundary. *Rev. Palaeobot. Palynol.* **1**, 141-150 (1967).

24. J. Pan. Terrestrial Devonian in South China. *Symp. on Devonian in South China*. Geological Publishing House, 240-269 (1978).

25. E. Paproth. The Devonian - Carboniferous boundary.*Lethaia* **13**.(4), 287 (1980).

26. G. Playford. Lower Carboniferous microfloras of Spitsbergen - Pt2 -*Palaeontology* **5**, 619-678. pls,88-95 (1963).

27. G. Playford. Plant microfossils from the Upper Devonian and Lower Carboniferous of the Canning Basin, Western Australia. *Palaeontographica.* **158B**, 1-71 (1976).

28. C.A. Sandberg, M. Streel and R.A. Scott. Comparison between conodont zonation and spore assemblage at the Devonian - Carboniferous boundary in the western and central United States and in Europe. *7 ième Cong.Int.Strat.Géol.Carb.,Krefeld,C.r.*1,179-203 (1972).

29. M. Streel. Distribution stratigraphique et géographique d' *Hymenozonotriletes lepidophytus* KEDO, d' *Hymenozonotriletes pusillites* KEDO et des assemblages tournaisian. - *Congrès et Colloques Univ. Liège,* **55**. 121-147 (1970).

30. E. Turnau. Spore zonation of uppermost Devonian and lower Carboniferous deposits of Western Pomerania (N.Poland). *Meded.Rijks.Geol. Dienst* **30**, 1-35 (1978).

31. Z. Wen and L. Lu. Devonian Carboniferous assemblage from Xiaomu section of Quannan, Jiangxi, China. *Acta Palaeontologica Sinica.* **32 (3)**, 303-320 (1993).

32. W. Yang, R. Neves and B. Liu. On the Occurrence of *Retispora lepidophyta* (Kedo) Playford in West Yunnan , S.W.China. In: Ren J S *et al* (ed) *Proceedings of 1st Intern. Symp. on Gondwana Dispersion and Asian Accretion (IGCP- Project. 321)* Wuhan: Press of China University of Geosciences, 22-25 (1991).

33. W. Yang. An investigation into Upper Palaeozoic palynology in the suture zone area (W.Yunnan, S.W.China) between Gondwana and Laurasia plates and its geological significance. unpublished Ph.D. thesis, a joint study between Sheffield University and China University of Geosciences. (1993).

34. Y. Yang. Spores. The late Devonian and Early Carboniferous strata and Palaeobiocoenosis of Hunan. Geological Publishing House. 51-52 (1987).

35. C. Yu (ed.). Devonian - Carboniferous Boundary in Nanbiancun, Guilin, China. Aspects and Records. Science Press, Beijing, China, (1988).

Proc. 30ᵗʰ Int'l. Geol. Congr., Vol. 12, pp. 108-114
Jin and Dineley (Eds)
© VSP 1997

Shallow-Marine Benthic Palaeoecology of the Late Pleistocene Paleo - Tokyo Bay: a Moderately Sheltered Bay Fringed with Barrier Islands

YASUO KONDO[1], TAKANOBU KAMATAKI[2] and YASUNORI MASAKI[1]
[1]*Department of Geology, Kochi University, Kochi, 780 Japan*
[2]*Department of Geology and Mineralogy, Kyoto University, Kyoto, 606 Japan*

Abstract

On the basis of previously established sedimentary facies analysis, the fossil composition and taphonomic features of the benthic fossil assemblages of the late Pleistocene Paleo-Tokyo Bay deposits (Kioroshi Formation) were analyzed. The bay deposits consist of various distinct sedimentary systems, including ocean beach open to the Pacific, beach in the bay, bay lagoon, flood-tidal delta and tidal inlet. The last three were found only in Paleo-Tokyo Bay, and not in modern Tokyo Bay and nearby coastal sea. These environments were formed associated with the inferred barrier islands in the eastern rim of the bay. The tidal inlet fill deposit contains dense shellbeds consisting mainly of the venerid bivalve *Ruditapes philippinarum*. Some of the individuals show evidence for upward escape during rapid sedimentation. In the west of the tidal inlet occurred a flood-tidal delta built into the bay. An inshore to offshore arrangement of molluscan fossil assemblages are preserved as a progradational sequence of the delta. *Chion kiusiuensis* and *Gomphina neastartoides* are common in the upper delta front, and higher-diversity molluscan assemblages occurred in the lower delta front environment. Particularly, *Mactra chinensis* were abundant on the physically unstable middle delta front environment. Also, a small colony of the deep-burrowing mactrid *Raeta pellicula* was found in the same environment. Mactrids, such as *Pseudocardium sachalinensis* and *Tresus keenae*, were also common in the bay shoreface environment in older Paleo-Tokyo Bay deposits. Such abundance of the mactrid bivalve is characteristic of the benthic associations in the Paleo- Tokyo Bay.

Keywords: barrier island, flood-tidal delta, tidal inlet, taphonomy, paleoecology, bivalve, Pleistocene, Paleo-Tokyo Bay, Shimosa Group, Kioroshi Formation

INTRODUCTION

In palaeoecological studies of Quaternary and late Neogene strata, an actualistic approach is commonly employed, because most of the species occurring in the fossil assemblages are still living and ecological information on the species can be directly applied. In the history of studies of the molluscan fossils of the Shimosa Group, most of the environmental interpretations have been based on the modern distribution of the species present in the fossil assemblages [e.g., 11-13, 17-19,]. However, it is obvious that inferences based solely on the ecological information about living species easily result in mere confirmation that "the past was the same as the present". To solve this problem, we need a method not anchored in the present to infer the paleoecology, even in geologically young deposits as in the Quaternary. To this purpose, taphofacies analysis, combined with sedimentary facies analysis, have been available as bases of determining environments and processes of deposition of fossiliferous strata.

The Pleistocene shallow embayment called Paleo-Tokyo Bay encompassed roughly the same area of the present-day Kanto plain in central Japan. The bay was shallow, moderately sheltered, fringed with barrier islands in the eastern rim [7]. A branch of the warm current, Kuroshio, is interpreted to have flowed into the bay from the south [4]. Bivalves were highly abundant in the bay. No similar environments are found in modern seas around the Japanese Islands.

In this study, taphonomy and paleoecology of bivalves and other megabenthic invertebrate associations are briefly described, based on the previously established sedimentary facies and sequence stratigraphic analysis [7, 9, 14-16]. Representative fossils and their mode of occurrence are outlined in this paper for each sedimentary system or environment.

GEOLOGIC, STRATIGRAPHIC AND PALEOENVIRONMENTAL SETTINGS

The Kanto basin is a depression situated at the junction of the Northeastern Honshu Arch and Southwestern Honshu Arch, attaining more than 100 Km across. The deep depression (Kazusa Basin) was gradually filled with clastic sediments and a much shallower embayment fringed with barrier islands, appeared by the middle Pleistocene. The shallow-bay deposits (Shimosa Group) consist of alternating, richly fossiliferous, shallow-marine, sandy sediments and thinner, non-marine, muddy deposits, and forms a cyclothemic sequence. In the Kisarazu area, on the east coast of Tokyo Bay, the cyclic sequence has been stratigraphically divided into the Jizodo, Yabu, Kami-izumi, Kiyokawa, Yokota and Kioroshi formations in upward sequence [22]. The Paleo-Tokyo Bay deposit studied in this paper is the Kioroshi Formation, which has been dated as forming during the last interglacial time.

Kondo [5] reconstructed the paleogeography and paleoceanographic condition of the bay during the last interglacial time, based on the bivalve fossil assemblages. The result shows that the fossil assemblages of the Paleo-Tokyo Bay deposit are characterized by mixed occurrence of warm Kuroshio-type and cold Oyashio-type molluscs with varying ratios. Generally, Oyashio-type molluscs are more abundant in the northern part of the bay, and open-coast molluscs of Kuroshio-type occur in the center of the bay. The paleoceanographic condition of the bay was therefore characterized by the strong influence of the Kuroshio from the south through the mouth of the modern Tokyo Bay, and there was a temperature gradient from south to north. In addition, another style of mixed occurrence of molluscs is recognized; open-coast molluscs and sheltered-bay-type molluscs occur together, although they are well separated in modern seas nearby. This is caused by the mixing of coastal water and oceanic water through the tidal inlet connecting the bay and the Pacific. The paleoceanographic condition of the Paleo-

Tokyo Bay was thus very complex, and no similar marine environments are found around the modern Japanese Islands.

RESULTS AND DISCUSSIONS

Flood-tidal Delta Taphofacies (Kioroshi delta)
The prograding delta deposit originally inferred for the Kioroshi Formation distributed in the Kioroshi area, Chiba Prefecture [14] was found to be a flood-tidal delta associated with a tidal inlet between the barrier islands inferred in the eastern rim of the bay [15]. Okazaki and Masuda [15] recognized fossiliferous, large-scale foreset beds dipping southwest and interpreted these as foreset delta beds. Kojima [1, 2] recognized three subunits in the sandy fossiliferous strata of the Kioroshi Formation, that is the foreset beds of the flood-tidal delta: densely fossiliferous sand including many species of bivalve (0-5 m), less fossiliferous, coarser-grained sand (0-6 m) and sand dollar debris bed (3-8m) in upward sequence. This sequence is interpreted as representing an offshore to inshore facies change on the delta front.

Molluscan fossil assemblages were analyzed for the delta foreset beds at Tsurumaki, Immba-mura, Chiba Prefecture. Numbers of individuals and of species per 5,000cm^3 sediments decrease upward (inshore), as is typified by *Scapharca subcrenata, Solen krusensterni, Saccella confusa, Nitidotellina nitidula, Raeta pulchellum, Cryptomya busoensis, Cycladicama semiasperoides, Callista chinensis, Semelangulus tokubeii, Oblimopa japonica* and *Anisocorbula venusta*. These are known to occur in the shallow subtidal to upper sublittoral zones. On the contrary, *Chion kiusiuensis* and *Gomphina neastartoides* are more abundant, or only found in the upper, or inshore samples. These species are known to inhabit the intertidal or very shallow subtidal zones. *Mactra chinensis* is very abundant in all the samples, with the maximum in the middle horizons. This may be consistent with its modern distribution in the lower intertidal to shallow subtidal zones. The stratigraphic changes in the composition, that is the inferred proximal to distal changes on foreset bottoms of the flood-tidal delta, are consistent with the modern distribution of molluscs. Also, these systematic changes mean that the fossil associations reflect, to some extent, those of the original community, although most of the specimens were not *in-situ*.

It is to be noted that the active, and shallow-burrowing bivalve, *Mactra chinensis* is abundant in physically unstable foreset bottoms of the flood-tidal delta systems. Most of the specimens are disarticulated, but broken or abraded specimens are uncommon. The conspicuously dense occurrence of this bivalve can be taken as indigenous, or buried within and around the habitat. A small colony of the well-inflated and thin-shelled mactrid *Raeta pellicula* was found on the delta front sand facies [3]. The predominant occurrence of mactrid bivalves in physically unstable sandy substrates is characteristic of the Paleo-Tokyo Bay and modern coastal sea in Japan.

Tidal Inlet Taphofacies (Tako Channel of Sato [21])
In the eastward, or seaward of the flood-tidal delta, a deep channel existed through the

barrier island open to the Pacific. The channel fill deposit is up to 20 m thick, consisting of cross-bedded gravelly sand including intraclasts, and silt including peat and plant debris in the lower unit, alternating beds of silt and very fine sand, massive silt, bioturbated, silty, very fine to medium sand containing *in-situ* molluscs, and densely fossiliferous coarser-grained sand in the upper unit [21]. The fossil assemblages in the tidal inlet system are dominated by the active-burrowing, venerid bivalve *Ruditapes philippinarum*, with subordinate elements of *Pillucina striata*, *Mactra nipponica*, *M. chinensis*, *Clinocardium californiense*. *Mactra chinensis* is locally very abundant in the tidal channel fill deposit, as well as the flood-tidal delta-front deposit.

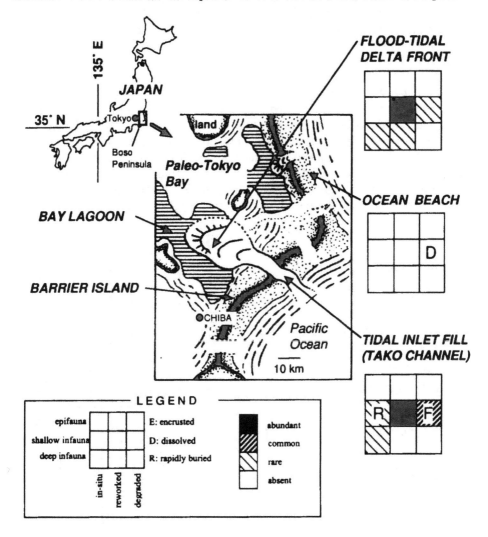

Figure 1. Bivalve taphofacies in the Paleo-Tokyo Bay.

In a fossiliferous channel fill sediments exposed at Asakura, mostly conjoined specimens are crowded and many of them show evidence of rapid burial. A fossil assemblage containing a *R. philippinarum* population formed by rapid burial is found (Fig. 2) only in this sedimentary system. Kranz [6] described an anterior-up, inverted orientation from the normal life position for most bivalves in experimental rapid burials. Many individuals of *R. philippinarum* at Asakura, are conjoined and oriented not randomly in the strata; the tendency of anterior-up is apparent. The insides of the conjoined specimens are not filled with sediment. *In-situ* preservation of *R. philippinarum* has not been reported, because it lives in physically unstable substrates where erosion is common, and the preservation potential is low. The unusually preserved *R. philippinarum* indicating rapid burial was probably caused by the higher sedimentation rate in the process of filling the tidal inlet during a period of rising sea level.

Ocean Beach to Shoreface Taphofacies
A typical facies of ocean beach to shoreface was reported from the Kioroshi Formation at Yamada, Kitaura-mura, Namegata-gun, Ibaraki Prefecture [8]. About a 10 m sequence consisting of bioturbated fine sand (lower shoreface), alternating beds of wave-rippled, coarse sediments and mud in the lower unit, fining-upward sets consisting of cross- bedded sand and hummocky cross-stratified sand in the middle, and alternating beds of megarippled gravelly sand and mud in the upper (middle to upper shoreface), and low- angle cross-beds and current-rippled sand (upper shoreface to

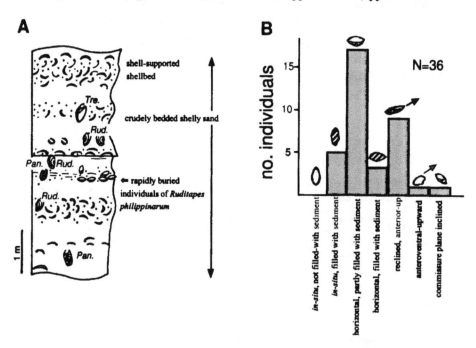

Figure 2. A: Columnar section, and B: the mode of occurrence of Ruditapes philippinarum in the fossiliferous tidal inlet fill deposit at Asakura, Shibayama-machi, Chiba Prefecture.

foreshore). Shell material is extremely rare, except for those associated with gravel in the middle unit.

Present-day foreshore of an exposed ocean beach in central Japan is densely populated by the donacid *Chion semigranosus*, but no fossil assemblages dominated by the donacid have ever been found. Shell material is completely dissolved out in almost all the beach facies examined. The absence is therefore the result of shell dissolution, in addition to the low preservation potential due to the frequent reworking. *Macaronichnus* burrows [10] preserved in beach and shallow subtidal sandy facies may be the only indication of activities of the benthos in the beach facies.

Acknowledgements

We thank Dr. H. Okazaki of Natural History Museum and Institute, Chiba, for helpful discussions, and Prof. A. J. Boucot of Oregon State University for reviewing the manuscript.

REFERENCES

1. N. Kojima. Geological study of the Kioroshi District, Chiba Prefecture, Japan - the studies on the Narita Group (1) , *Jour. Geol. Soc. Japan* 64, 165-171 (1958).
2. N. Kojima. On the mode of occurrence of the fossil shells in the Narita Formation of the Kioroshi district, Chiba Prefecture, Japan. *Jour. Geol. Soc. Japan* 64, 213-221 (1958).
3. Y. Kondo. Preserved life orientations of soft-bottom infaunal bivalves: documentation of some Quaternary forms from Chiba, Japan, *Nat. Hist. Res.(Nat. His. Mus. Inst., Chiba)* 1, 31-42 (1991).
4. Y. Kondo. An open-coast shallow-marine molluscan fossil assemblage from the Late Pleistocene of Matsudo, Chiba: Implication for paleoceanographic reconstruction of Paleo-Tokyo Bay in the Last Interglacial, *Jour. Nat. Hist. Mus. and Inst., Chiba* 2, 1-8 (1991).
5. Y. Kondo. Paleoceanography and paleogeography of Paleo-Tokyo Bay in the Last Interglacial: A reconstruction based on the bivalve fossil assemblages, *Progr. Abst., Japan Ass., Quat. Res.*, 21: 102-103 (1991).
6. P. M. Kranz. The anastrophic burial of bivalves and its paleoecologic significance, *Jour. Geol.* 82, 237-265 (1974).
7. Y. Makino and F. Masuda (Eds). Barrier islands of the Paleo-Tokyo Bay, In: *Field Excursion Guidebook, 96th Ann. Conf. Geol. Soc. Japan*, 151-199. Geol. Soc. Japan (1989).
8. Y. Makino, I. Yoshikawa, N. Terakado and Y. Katsura. Megaripple and storm deposits, In: Barrier islands of the Paleo-Tokyo Bay. Y. Makino and F. Masuda (Eds). 160-163. *Field Excursion Guidebook, 96th Ann. Conf. Geol Soc. Japan*. Geol. Soc. Japan (1989).
9. N. Murakoshi and F. Masuda. Estuarine, barrier-island to strand plain sequence and related ravinement surface developed during the last interglacial in the Pleo-Tokyo Bay, Japan, *Sediment. Geol.* 80, 167-184 (1992).
10. M. Nara. What is the producer of "trace fossil of Excirolana chiltoni" ? - tracemaking mechanism of Macaronichnus segregatis, *Kaseki (Fossil)* 56, 9-20 (1994).
11. S. O'Hara. Molluscan fossils and constituent minerals of the Narita Formation, *Jour. Coll. Arts Sci., Chiba Univ.* B 4, 49-78 (1971).

12. S. O'Hara. Molluscan fossils from the Kami-izumi formation (s.l.), *Jour. Coll. Arts Sci., Chiba Univ.* **B 11**, 59-89 (1978).

13. S. O'Hara. Molluscan fossils from the Shimosa Group (1. Yabu and Jizodo Formations of the Makuta district), *Jour. Coll. Arts Sci., Chiba Univ.* **B 15**, 27-56 (1982).

14. H. Okazaki and F. Masuda. Two types of prograding deltaic sequence developed in the late Pleistocene Paleo-Tokyo Bay, *Ann. Rep., Inst. Geosci., Univ. Tsukuba* **9**, 56-60 (1983).

15. H. Okazaki and F. Masuda. Arcuate and bird's foot deltas in the late Pleistocene Palaeo-Tokyo bay, In: M. K. G. Whateley and K. T. Pickering (Eds). 129-138. *Deltas: sites and traps for fossil fuels.* Geol. Soc. London, Spec. Pub. **35**, (1988).

16. H. Okazaki and F. Masuda. Depositional systems of the Late Pleistocene sediments in Paleo-Tokyo Bay area, *Jour. Geol. Soc. Japan* **98**, 235-258 (1992).

17. K. Oyama. Fossil communities of the coastal water (No. 1), *Mis. Rep. Res. Inst. Nat, Res.* **31**, 54-59 (1953a).

18. K. Oyama. Fossil communities of the oceanic water (No. 1), *Mis. Rep. Res. Inst. Nat, Res.* **32**, 23-31 (1953b).

19. K. Oyama and S. Ishiyama. A problem for estimating sedimentary condition - as an example of horizons in Kami-izumi, Chiba Prefecture, *Bull. Geol. Surv. Japan* **19**, 569-574 (1968).

20. Y. Saito. Morphology and sediments of the Obitsu Delta in Tokyo Bay, Japan, *Jour. Sed. Soc. Japan* **35**, 41-48 (1991).

21. H. Sato. Stratigraphy of the middle-upper Pleistocene Shimosa Group distributed in the area from Naruto Town , Sanbu-gun to Yoka-ichiba City, Chiba Prefecture, *Jour. Nat. Hist. Mus. and Inst., Chiba* **2**, 99-113 (1991).

22. S. Tokuhashi and H. Endo. *Geology of the Anesaki district Quadrangle Series, scale 1: 50,000*, Geol. Surv. Japan, 136 p. (1984).

Proc. 30ᵗʰ Int'l. Geol. Congr., Vol. 12, pp. 115-126
Jin and Dineley (Eds)
© VSP 1997

Freshwater Molluscan Fauna of the Miocene-Pliocene Churia (Siwalik) Group of Nepal and Their Palaeoecological Implication

Damayanti Gurung[1] Katsumi Takayasu[2] and Keiji Matsuoka[3]

[1] *Graduate School of Science and Technology, Niigata University, Niigata 950-21, Japan.*
[2] *Research Center for Coastal Lagoon Environments, Shimane University, Matsue 690, Japan.*
[3] *Toyohashi Museum of Natural History, Toyohashi 441-31, Japan.*

Abstract

Freshwater molluscan fossils occur abundantly in the Churia Group, in West-central Nepal. Study to evaluate their faunal composition and stratigraphical distribution shows some interesting trends. The molluscan fossils are found to occur most abundantly between ca. 10.5Ma and ca. 3.5Ma, before and after which they are rare. Relative abundance of localities as well as diversity increases toward the upper part with the fossil localities concentrated along four horizons. The faunal composition shows two phases of change, around 7 Ma and 4 Ma, with disappearance and appearance of taxa.

Keywords: Fossil, Freshwater molluscs, Churia, Siwaliks, Miocene, Pliocene, Nepal.

INTRODUCTION

The Middle Miocene - Pleistocene Churia Group (equivalent to the Siwaliks of India and Pakistan) of Nepal is distributed along the southern foothills of the Nepal Himalayas. It is composed of a thick molasse sedimentary sequence and is nearly 6000 m thick with overall coarsening upward succession of alternating siltstone, sandstone and conglomerates (15; 17; 22; 27). Diverse vertebrate fauna and flora along with invertebrate molluscan fauna have been reported from these sedimentary deposits (5; 12; 18; 24; 26; 33; 34).

Aquatic molluscan specimens are generally found to be preserved with very little transportation, and because of their habitat preference, distribution is restricted for many taxa (6). Therefore, change in their stratigraphic distribution and faunal composition may provide evidence for inferring ecology and climate changes as well as change in the river connections. In the Nepal Siwaliks freshwater molluscan faunas occur in relative abundance compared to other groups. Studies related to it are few (26; 33). Most of the palaeontological studies related to the freshwater molluscs from the Siwaliks were carried out in Burma (4; 32), India (3; 8; 9; 18) and Pakistan (7; 30; 31). Only recently has description of the bivalve fossils from Nepal Siwaliks been carried out by Takayasu *et al.* (26) with identification of 9 species belonging to 4 genera. Gurung *et al.* (in prep.) have identified 14 taxa of gastropoda belonging to 9 genera from the same area.

Location and Geological Setting

The study area lies about 250 km southwest of Kathmandu. The area includes the valleys of the Arung, Binai and Tinau Khola (Fig.1). The specimens were collected from localities distributed along these valleys. The Nepal Siwalik is well studied in this area. Detailed geological mapping (Fig.1) with the establishment of the lithostratigraphy and palaeomagnetic data has been carried out (14; 17; 27-29).

According to Tokuoka *et al.* (27), the Group is divided into the Arung Khola, Binai Khola, Chitwan and Deorali Formations in ascending order. The ages of the formational boundaries followed here are on magnetic polarity correlated to those of Cande and Kent (11) (Fig.2). The Arung Khola Formation is composed of alternating beds of variegated mudstone and fine sandstone, and is further divided into lower, middle and upper members. The age of the Arung Khola Formation is found to be from 9.5-14 Ma. The Binai Khola Formation consists of alternating beds of coarse 'pepper and salt' sandstone and mudstone, and is also divided into lower, middle and upper members. The age of this Formation ranges from 2.5-9.5 Ma. The Chitwan Formation is mainly composed of well sorted conglomerates with age from 1.3-2.5 Ma. The upper most Deorali Formation consists of ill-sorted boulder conglomerates, deposited from 1.3 Ma onward. The Group is in tectonic contact with the northern older Midland Group, and southern, younger, recent alluvium. Furthermore, the Group is divided into a south belt and north belt by the Central Churia Thrust.

Molluscan Fauna

There are more than 45 localities (Fig.1) distributed in the above-mentioned area, however, not all yield well-preserved specimens. Few localities contain shell fragments in the channel fill and one locality (F21) is composed of large blocks of lag beds. Most of the localities considered here yield well-preserved, mostly intact shells, which could be considered as indigenous specimens, although, some form of deformation is present.

At most localities mollusca are predominant in the fossil assemblage which generally consists of Charophyte gyrogonites, carbonized plant fragments, fish scales and pharyngeal teeth, crocodile teeth and bone fragments. At some localities only molluscan fossils are recovered. They are usually found at the top of fining-upward cycles within the large coarsening-upward sequence, mostly within mud-dominated sequences.

Stratigraphically the fauna is distributed from the upper member of the Arung Khola Formation to the middle member of the Binai Khola Formation. Many fossil localities are reported by Tokuoka *et al.* (27-29) with many added during subsequent field work

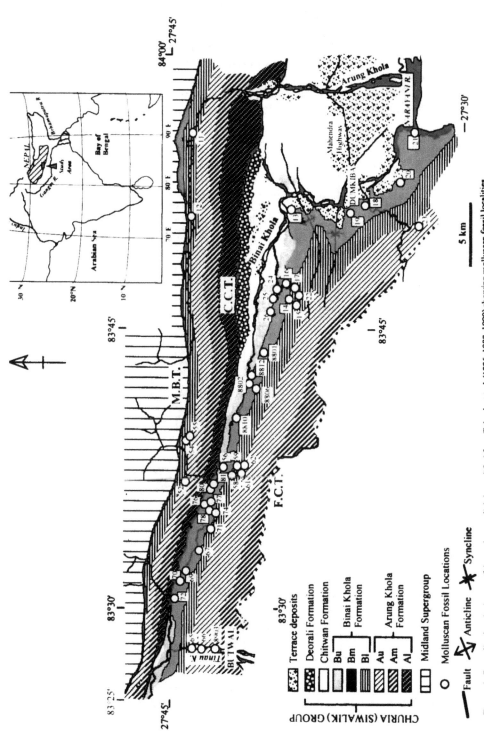

Figure 1. Generalized geologic map of the study area slightly modified from Tokuoka et al. (1986, 1988, 1990) showing molluscan fossil localities

by the authors. Many fossil localities are located in the middle member of the Binai Khola Formation in the south belt.

It is difficult to correlate precisely each fossil location laterally, as the fossil-bearing beds are not continuously exposed. However, the fossil localities are distributed in rather concentrated horizons, stratigaphically and are loosely grouped into four Fossil Horizons on their stratigraphic position (Fig.2). They are as follows, Au, Bl, Bm-1 and Bm-2 Fossil Horizons, in ascending order. The molluscan taxa present in these horizons

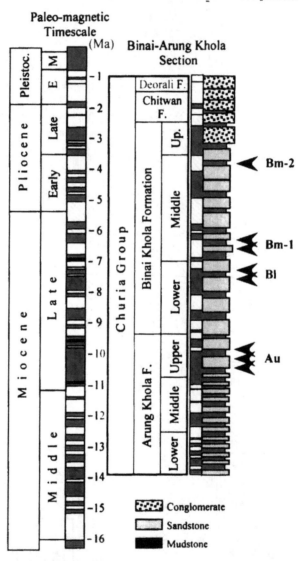

Figure 2. Simplified stratigraphy of the Churia Group with polarity compared to palaeomagnetic timescale of Cande and Kent (1994). The approximate stratigraphic position of the fossil horizons are as shown.

are given in table 1. Most of the taxa so far reported from these localities are shown in figs.3 and 4.

Fossil Horizon and Faunal composition
Au Fossil Horizon
This horizon consists of five fossil localities, F-65, F-66, F-8903, F-52 and F-23, within the upper member of the Arung Khola Formation. Specimens with completely preserved shells are recovered from the two uppermost localities (F-66 and F-65) only. The lower three localities yield comparatively poorly preserved shells. The faunal composition is mainly dominated by Prosobranchia *Bellamya* sp. A or by Unionidae *Parreysia binaiensis* with *Melanoides* sp., *Brotia palaeocostula, Brotia* sp. C and Unionidae *Lamellidens arungensis, P. zigzagicostata* in association.

Bl Fossil Horizon
Compared with the lower Au horizon, more localities with more well-preserved specimens are found in the Bl Fossil Horizon. The fossil localities (F-11, 12, 15, 22, 61, 55, 54, 74, 73, 59) are found concentrated toward the upper part of the lower member of the Binai Khola Formation. Faunal composition is similar to the lower one, however, with fauna mostly dominated by Unionidae *Parreysia binaiensis* with Prosobranchia *Bellamya* sp. A, *Melanoides* sp., *Brotia palaeocostula, Brotia* sp. C and Unionidae *Lamellidens arungensis, P. zigzagicostata.* in association.

Bm-1 Fossil Horizon
This horizon is found along the lower part of the middle member of the Binai Khola Formation. It consist of 7 localities (F-14, 17, 67, 71, 79, 60, 8806). The fossil fauna is mainly dominated by Prosobranchia *Brotia palaeocostula, Brotia* sp. B, *Brotia* sp. C with *Bellamya celsispiralis, Melanoides* sp., *Lamellidens arungensis, L. longiformis, Indonaia churia,* in association.

Bm-2 Fossil Horizon
These localities are distributed in the upper part of the middle member of the Binai Khola Formation. This uppermost horizon has the highest number of fossil localities (F-21, 20, 18, 19, 13, 16, 24, 25, 26, 8801, 8810, 88012, 72) and with relatively high yield. The Bm-2 Horizon fauna is composed of many molluscan faunal compositions. Most commonly they are dominated by Viviparidae *Bellamya celsispiralis, Bithynia* sp., Unionidae *Physunio chitwanensis, Lamellidens arungensis,* and Thiaridae *Brotia palaeocostula.* Generally operculat *Pila* sp., Pulmonata such as *Indoplanorbis* sp., *Gyraulus* sp., *Lymnaea* sp., Unionidae *Parreysia* sp., *Indonaia churia, Indonaia narayani, Indonaia jimuriensis, Indonaia tenella,* small clams Pisidiidae *Pisidium* sp. are found in association.

DISCUSSION

The molluscan fauna from the Nepal Siwalik belongs to genera of the Viviparidae, Ampullariidae, Thiaridae, Bithyniidae, Lymnaeidae, Planorbidae, Unionidae,

Pisidiidae, that are aquatic in habit. The occurrence of these fossils in this Group in the west-central part shows some interesting trends. Most aquatic molluscs have restricted distribution ability, excluding small-sized ones and some pulmonates which have a high possibility of passive dispersal. Geographical distribution is controlled by the freshwater bodies and their connecting routes (13; 25). Large freshwater mussels (Unionacea), many families of prosobranchiates and many pulmonates are mainly dispersed by freshwater routes (6). Because of this , these groups of freshwater molluscs are considered important for palaeobiogeographic reconstruction (6). Their occurrence also depends on the availability of suitable environment. The Nepal Siwalik molluscan fauna belongs to the genera which generally inhabit large rivers with backmarshes, swamps, floodplain lakes, generally shallow slow-flowing water. The abundant occurrence of molluscan fauna along certain horizons in comparison to the other fauna indicates that such habitats were extensively developed at certain times.

Table 1. The distribution of the freshwater molluscan taxa along the fossil horizons within the Churia Group.

Taxa	Fossil Horizons			
	Au	Bl	Bm-1	Bm-2
Gastropoda				
Bellamya celsispiralis			+	+
Bellamya sp. A	+	+		
Bellamya sp. B			+	+
Angulyagra sp.				+
Bithynia sp.	+	+	+	+
Pila sp.				+
Melanoides sp.	+	+	+	+
Brotia palaeocostula	+	+	+	+
Brotia sp. A				+
Brotia sp. B			+	+
Brotia sp. C	+	+	+	+
Lymnaea sp.			+	+
Indoplanorbis sp.				+
Gyraulus sp.				+
Bivalvia				
Lamellidens arungensis	+	+	+	+
L. longiformis				+
Indonaia churiua			+	+
Indonaia narayani			+	+
Indonaia tenella			+	+
Indonaia jimuriensis				+
Parreysia binaiensis	+	+		
P. zigzagicostata	+	+		
Physunio chitwaensis				+
Pisidium sp.				+

As mentioned above, the molluscan fossils mainly occur at the top of fining-upward cycles, in fine siltstone, occasionally in fine sandstone. It is believed that the fine member of the fining-upward cycles generally includes strata formed on floodplains flanking active stream channels (1). At some sections many thick fining-upward cycles are without molluscan fossils. Besides that the occurrence of the fossils is mainly concentrated along four fossil horizon. This probably indicates that suitable habitat was not extensively developed at all depositional intervals.

The first occurrence of molluscan fossils coincides with the predominance of sheet splay deposits from the upper part of the Arung Khola Formation, indicating the prevalence of a meandering river system with extensive development of floodplains (16). The presence of such an environment is also indicated by the vertebrate fauna from the Nepal Siwalik, which is largely aquatic, represented by fish, turtles, crocodiles and snakes (18).

At first occurrence faunal composition is comparatively simple with a smaller number of genera in the lower horizon. The faunal composition does not show change from the first occurrence until about 7 Ma, from the lower part of the middle member of the Binai Khola Formation, when it changes with addition of such taxa as *Lymnaea* sp., *Indonaia* sp., Pisidium sp. and disappearance of others like *Parreysia binaiensis*, *P. zigzagicostata*, and *Bellamya* sp. According to Quade *et al.* (20; 21), the period around 7 Ma is associated with climatic changes, intensification of monsoon and change in flora from largely C3 vegetation to C4 grassland. The depositing river system is also considered to be larger than during the deposition of the lower part (21). The appearance of the small molluscs like *Lymnaea* sp. and *Pisidium* sp., which could be passively distributed by animals probably associated with the migration of the land animals triggered by the change in climate. The disappearance of the Unionidae *Parreysia* sp. may probably be associated with change in climate with related unfavorable changes in habitat.

The other major change in the molluscan faunal composition is observed around 4 Ma, with the addition of many genera, like *Bellamya celsispiralis*, *Angulyagra* sp., *Gyraulus* sp., *Indoplanorbis* sp., *Lamellidens longiformis*, *Parreysia* sp., *Physunio chitwensis*. Although, the molluscan faunal composition has genera that are at present with wide distribution like *Melanoides* sp., *Bithynia* sp., *Pisidium* sp., many are large Unionidae and gastropods that have much restricted distribution as mentioned above. Among them the appearance of Sino-Indian elements like *Angulyagra* sp. and *Physunio* sp. (23) is interesting to note. Recent distribution of the species of these genera is in Assam, Burma, Thailand, South China (10; 23) and they generally inhabit stagnant water bodies (2). The presence of *Charophytes gyrogonites*, which are considered a stagnant water element, with most molluscs also indicates shallow stagnant water bodies. At about the same time, change to swampy condition is indicated by pollen analysis conducted in the Surai Khola Section of the western Nepal Siwaliks (24). It seems that during a short depositional interval, just before the progression of the coarse alluvial fan deposit of the upper Chitwan and Deorali Formations, there was a period with extensive development of stagnant water bodies.

CONCLUSION

Freshwater molluscan fossils are recorded from the upper member of the Arung Khola Formation, from about 10.5 Ma, up to the middle member of the Binai Khola Formation, about 3.5 Ma. The fossil localities seem to be concentrated along four horizons, Au, Bl, Bm1 and Bm2, with increase in number of localities in the upper part, especially in the middle member of the Binai Khola Formation. Fossil freshwater molluscs occur extensively in the Binai Khola area as compared to the lower ones.

At first occurrence, at ca. 10.5 Ma, the molluscan faunal composition consists of prosobranchia *Bellamya* sp., and Unionidae, *Lamellidens* sp., *Parreysia* spp. The faunal composition shows two distinct phases of change further up the section, at two intervals, at ca. 7Ma and at ca. 4Ma.

The first change in faunal composition is observed at ca. 7 Ma, in the Bm-1 horizon fauna. The dominant species of the Au and Bl horizons, Viviparidae *Bellamya* sp. A and Unionidae *P. binaiensis* and *P. zigzagicostata* are not found in the Bm-1 and Bm-2 horizon fauna. The Bm-1 fauna is represented by species of Thiaridae *Brotia* and Unionidae *Lamellidens*, with the first appearance of genus *Indonaia*, and genus *Pisidium*. The faunal change seem to be associated with the change in climate and the associated environments.

The second change in the faunal composition is observed at ca. 4 Ma, with the addition of numerous taxa and diversification of others, in the fauna of the upper most horizon Bm-2. The Sino-Indian element like *Angulyagra* sp., *Physunio* sp. become quite dominant at few localities of this horizon. Together with many large Unionidae, small pulmonate gastropods like *Gyraulus, Indoplanorbis, Lymnaea* occur abundantly in this uppermost horizon. The operculate Prosobranchia Pila also appears. The diversification of the fauna in the upper part strongly indicates a period with extensive development of stagnant water environment during that depositional interval.

The present study shows the molluscan faunal change to be consistent with the regional changes taking place during Late Miocene to Early Pliocene in the Nepal Siwalik. Further study is needed to make clear their biostratigraphic, palaeobiogeographic and ecological significance.

Acknowledgments

We acknowledge the support and assistance provided by Prof. K. Kizaki (Professor Emeritus of The University of Ryukyus), Prof. I. Kobayashi of Niigata University and the staff of the Department of Geology, Tribhuvan University. We are grateful to Prof. A. J. Boucot, Oregon State University for reviewing the manuscript.

REFERENCES

1. J.R.L. Allen. Fining-upwards cycles in alluvial successions, *Geol. J.* **4-2**, 229-246 (1965).
2. N. Annandale. Aquatic molluscs of the Inle lake and connected waters, *Rec. Indian Mus.* **14**, 103-182, pls.10-19 (1918).
3. N. Annandale. Indian fossil viviparae, *Rec. Geol. Surv. India* **51**, 362-367, pl.11 (1921).
4. N. Annandale. Fossil molluscs from the Oil-measures of the Dawana Hills, Tenasserim, *Rec. Geol. Surv. India* **55**, 1-6 (1924).
5. N. Awasthi and M. Prasad. Siwalik plant fossils from Surai Khola area, western Nepal, In : Jain and Tiwari (eds.) - Proc. Symp. 'Vistas in Indian Palaeobotany', *Palaeobotanist*, **38**, 298-318 (1990).
6. P. Banarescu. *Zoogeography of freshwaters, vol.1*, General distribution and dispersal of freshwater animals, Aula-Verlag, Wiesbaden (1990).
7. W.T. Blanford. Geological notes on the Hills in the neighborhood of the Sind and Punjab Frontier between Quetta and Dera Ghazi Khan, *Mem. Geol. Surv. India* **20 (2)**, 105-240, pl. 1-3 (1883).
8. S.B. Bhatia and A.K. Mathur. Some upper Siwalik and Late Pleistocene molluscs from Punjab, *Him. Geol.* **3**, 24-58 (1973).
9. S.B. Bhatia. Some Pleistocene molluscs from Kashmir, India, *Him. Geol.* **4**, 371-393 (1974).
10. R.A.M. Brandt. The non-marine aquatic mollusca of Thailand, *Archiv. Molluskenk* **105** (1974).
11. C. Steven Cande and D.V. Kent. A New Geomagnetic Polarity time scale for the Late Cretaceous and Cenozoic, *J. Geophys. Res.* **97-13**, 917-953 (1992).
12. G. Corvinus. The Surai Khola and Rato Khola fossiliferous sequences in the Siwalik Group, Nepal, *Him. Geol.* **15**, 49-61 (1994).
13. G.M. Davis. Historical and ecological factors in the evolution, adaptive radiation, and biogeography of freshwater mollusks, *Amer. Zool.* **22**, 375-395 (1982).
14. P. Gautam and E. Appel. Magnetic-polarity stratigraphy of Siwalik sediments of Tinau Khola section in West Central Nepal, revisited, *Geophys. J. Int'l.* **117**, 223-234 (1994).
15. K.W. Glennie and M.A. Ziegler. *The Siwalik formation of Nepal*, Int'l. 22 Geol. Congr. **15**, 82-95 (1964).
16. K. Hisatomi and S. Tanaka. Climatic and environmental changes at 97.5 Ma in the Churia (Siwalik) Group, west central Nepal, *Him. Geol.* **15**, 161-180 (1994).
17. K. Kizaki. An outline of the Himalayan Upheaval - A case study of the Nepal Himalayas, Japan International Cooperation (1994).
18. J. Munthe, B. Dongol, J.H. Hutchison, W.F. Kean, K. Munthe and R.M. West. New fossil discoveries from the Miocene of Nepal include a hominoid, *Nature* **303**, 331-333 (1983).
19. B. Prashad. On some fossil Indian Unionidae, *Rec. Geol. Surv. India* **60**, 308-312, pl. 25 (1927).
20. J. Quade, T.E. Cerling and J.R. Bowman. Development of the Asian monsoon revealed by marked ecologic shift in the latest Miocene of northern Pakistan, *Nature* **342**, 163-166 (1989).
21. J. Quade, J.M.L. Carter, T.P. Ojha, J. Adam and T.M. Harrison. Late Miocene environmental change in Nepal and the northern Indian subcontinent: Stable isotopic evidence from paleosols, *Geol. Soc. America Bull.* **107 (12)**, 1381-1397 (1995).
22. C.K. Sharma. *Geology of the Nepal Himalaya and adjacent countries*, Education Enterprise (P) Ltd., Kath., Nepal (1977).
23. N.V. Subba Rao. *Handbook freshwater molluscs of India*, Zool. Surv. India, Calcutta, India (1989).

24. Samir Sarkar. Siwalik pollen succession from Suria Khola of western Nepal and its reflection on paleoecology, In : K. P. Jain and R. S. Tiwari (eds.) - Proc. Symp. *Vista in Indian Palaeobotany', *Palaeobotanist* **38**, 319-324 (1990).

25. D.W. Taylor. Aspects of freshwater mollusc ecological biogeography, *Palaeogeogr. Palaeocliml. Palaeoecol.* **62**, 511-576 (1988).

26. K. Takayasu, D. Gurung and K. Matsuoka. Some new species of freshwater bivalves from the Mio-Pliocene Churia Group, west-central Nepal, Trans. *Proc. Palaeont. Soc. Japan* **179**, 157-168, 5 figs. (1995).

27. T. Tokuoka, K. Takayasu, M. Yoshida and K. Hisatomi. The Churia (Siwalik) Group of the Arung Khola Area, west-central Nepal, *Mem. Fac. Sci. Shimane Univ.* **20**, 135-210 (1986).

28. T. Tokuoka, S. Takeda, M. Yoshida and B.N. Upreti. The Churia (Siwalik) Group in the Western part of the Arung Khola Area, west-central Nepal, *Mem. Fac. Sci. Shimane Univ.* **22**, 131-140 (1988).

29. T. Tokuoka, K. Takayasu, K. Hisatomi, H. Yamasaki, S. Tanaka, M. Konomatsu, R.B. Sah and S.M. Rai. Stratigraphy and geological structures of the Churia (Siwalik) Group in the Tinau Khola-Binai Khola Area, west-central Nepal, *Mem. Fac. Sci. Shimane Univ.* **24**, 71-88 (1990).

30. H.F. Vokes. Unionidae of the Siwalik Series, *Mem. Connecticut Acad.* **9**, 37-48, pl. 3 (1935).

31. H.F. Vokes. Siwalik Unionidae from the collection of the second Yale North India Expedition, *Quart. J. Geol. Min. Met. Soc. India* **8**, 133-144, pl. 10 (1936).

32. E. Vredenburg and B. Prashad. Unionidae from the Miocene of Burma, *Rec. Geol. Surv. India* **61**, 371-374, pl. 12 (1921).

33. R.M. West, J. Munthe, J.R. Lukacs and T.B. Shrestha. Fossil mollusca from the Siwalik of Eastern Nepal, *Current Science* **44(14)**, 497-498 (1975).

34. R.M.West, J.H. Hutchison and J. Munthe. Miocene vertebrates from the Siwalik Group, western Nepal, *J. Vertebrate Paleo.* **11(1)**, 108-129 (1991).

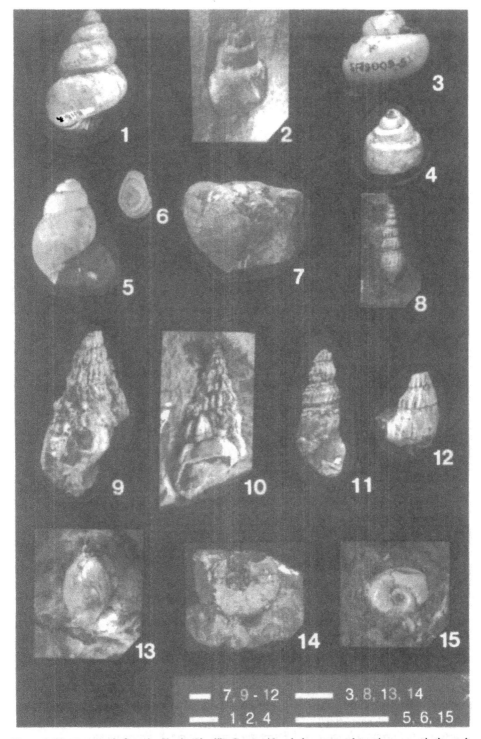

Figure 3. The Gastropoda from the Churia (Siwalik) Group with polarity compared to palaeomagnetic timescale of Cande and Kent (1994). The approximate stratigraphic position of the fossil horizons are as shown.

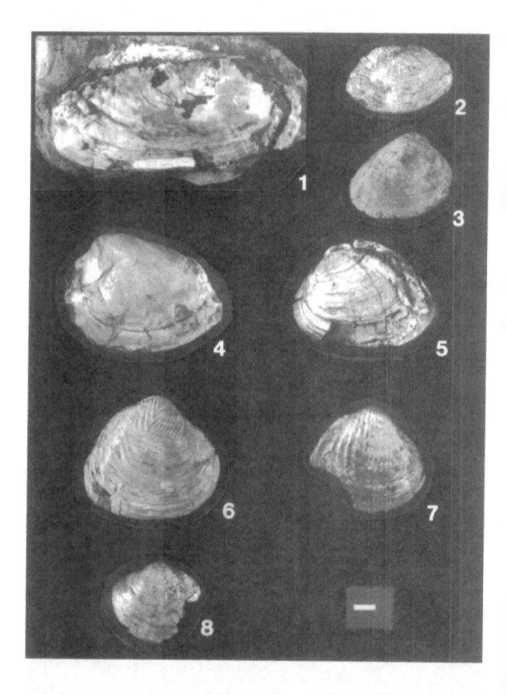

Figure 4. The Bivalvia from the Churia (Siwalik) Group of west-central Nepal. 1. *Lamellidens arungensis*, 2. *Indonaia tenella*, 3. *Indonaia narayani*, 4. *Indonaia churia*, 5. *Indonaia jimuriensis*, 6. *Parreysia binaiensis*, 7. *Parreysia zigzagicostata*, 8. *Parreysia* sp.

Proc. 30ᵗʰ Int'l. Geol. Congr., Vol. 12, pp. 127-135
Jin and Dineley (Eds)
© VSP 1997

The Significance of Freshwater Malacofauna of the Churia (Siwalik) Group in Nepal and the Himalayan Upheaval

KATSUMI TAKAYASU[1], DAMAYANTI GURUNG[2], and KEIJI MATSUOKA[3]

[1] *Research Center for Coastal Lagoon Environments, Shimane University, Matsue 690, Japan*
[2] *Graduate School of Science and Technology, Niigata University, Niigata 950-21, Japan*
[3] *Toyohashi Museum of Natural History, Toyohashi 441-31, Japan*

Abstract

The mode of occurrence and the stratigraphical record of fossil freshwater molluscs of the Middle Miocene to Pliocene Churia Group in Nepal (mostly equivalent to the Siwalik Group in Pakistan) provide much information about the paleoenvrionmental change related to the upheaval of the Himalayas. In the Churia sedimentary sequence, freshwater molluscan fossils are found from about 10.5 Ma to 3.5 Ma with faunal enrichment marked about 7 Ma [7]. Before 7 Ma, the fauna is rather simple and is mainly composed of typical Indian subcontinent unionid bivalves. After 7 Ma, the diversity of the fauna becomes high by addition of some species regarded as palearctic and Southeast Asian elements. The diversity rapidly increased after 4 Ma.

On the other hand, the first phase of the upheaval of the Himalayas reached to the acme from 9 to 11 Ma, and from 7 to 8 Ma became to have a great influence on the development of the monsoonal condition [1; 10]. The first occurrence of freshwater molluscs coincides with the development of flood deposits of the meandering river system caused by the seasonal heavy rain of the monsoon [8]. The introduction and enrichment of palearctic and Southeast Asian elements in the malacofauna after 7 Ma can be considered as transported by mammals and birds, which has been provoked by the formation of conspicuous climatic zonation in the Eurasian Continent. The major vegetational change at 6 to 7 Ma in the flood plain from forest to grassland, as indicated by the stable isotopic data of pedogenic carbonate, also provides the preferable condition of animal migration along the rivers.

Keyword: freshwater molluscs, Churia Group, Nepal Siwaliks, Miocene, Pliocene, Himalayan upheaval, environmental change.

INTRODUCTION

The upheaval of the Himalayas as well as that of the Tibetan Plateau is considered to have affected the climate and the distribution of the land biota of the Asian Continent. Many studies of the Siwaliks, as a molasse of the Himalayas, have made clear the process providing much information from the geological, paleontological and biological points of view. Paleontological work, however, had a tendency to focus mainly on vertebrate fossils in connection with the evolution of Primates, other taxa have been regarded as making a rather minor contribution .

The previous works on the molluscan fossils from the Siwaliks are very few and most were published before 1945 [2; 11; 12; 13; 14; 15; 16; 24; 25; etc.]. After that, only several supplemental but valuable reports were published [4; 26]. Recently we published a descriptive paper on bivalves from the Nepal Siwaliks [19] and the sequel on the gastropods is to be published in the future.

Though some species are still indeterminate because of the paucity of previous work on the fossils as well as the living taxa, we will discuss in this paper the relationship between the geological distribution of the Siwalik freshwater molluscs and the process of the Himalayan upheaval and related phenomena.

OUTLINE OF THE STRATIGRAPHICAL DISTRIBUTION OF THE FRESHWATER MOLLUSCS

The stratigraphy of the study area (fig.1) was established by Tokuoka *et al.* [19; 20; 21]. According to them, the Churia Group (mostly equivalent to the Siwalik Group in Pakistan) is divided into the Arung Khola, Binai Khola, Chitwan and Deorali Formations in ascending order (fig.1). The former two formations are subdivided lithologically into lower, middle and upper members. Their ages are determined by paleomagnetic measurement. According to the recent geomagnetic polarity timescale [5], the deposition of the Group started from before 14 Ma and continued at least upto 1.3 Ma.

Molluscan fossils range from the upper member of the Arung Khola Formation to the middle member of the Binai Khola Formation. The characteristics of the fauna will be described in detail in a companion paper [7]. The stratigraphical distribution is concentrated into four horizons; Fossil Horizons Au (ca. 10.5 Ma), Bl (ca.7.5 Ma), Bm1 (ca. 6.5 Ma) and Bm2 (ca. 3.5Ma). The number of fossil localities increases upward and at the uppermost, Fossil Horizon Bm2, the highest concentration of fossil localities occurs. Fossils are commonly found in the dark gray-colored siltstone of the uppermost part of the fining-upward sequence and occur mostly *in situ*.

From the study area, we discriminated 14 species of gastropods and 10 species of bivalves including indeterminate species. The faunas of the lower Fossil Horizons, Au and B1, show rather simple assemblages composed of *Bellamya* sp. A, *Melanoides* sp., *Brotia palaecostula*, *Brotia* sp., *Lamellidens arungensis*, *Parreysia binaiensis* and *P. zigzagicostata*. The diversity of the species increases upwards in proportion to the number of fossil localities, and, except for two species of *Parreysia* and one species of *Bellamya*, almost all species occur abundantly at the Horizon Bm2(fig.2).

The main change of faunal composition is recognized at about 7 Ma, between B1 (ca.6.5 Ma) and Bm1 (ca.7.5 Ma) Horizons. Although the both Horizons are rather near each other stratigraphically, two bivalve species of *Parreysia* which are dominated before 6.5 Ma disappear and are replaced by species of *Indonaia* and *Physunio* from Bm1 Horizon onward. The enrichment after 7.5 Ma is not only of these large-sized molluscs but also of such small-sized molluscs as *Pisidium* sp. and *Bithynia* sp. Though the opercula of the latter species have been found from the Au Horizon, their abundant occurrence is recognized at the Bm2 Horizon (ca. 3.5 Ma). Besides, gastropods such as *Angulyagura* sp., *Lynnaea* sp. and two planorbids of *Indoplanorbis* sp. and *Gyraulus* sp. contribute to faunal composition at that time.

ENVIRONMENTAL CHANGE FROM THE DATA OF SEDIMENTOLOGY AND STABLE ISOTOPE ANALYSIS

According to the Ocean Drilling Program (ODP) reports about the Bengal Fan, the uplift of the Himalayas started from 17.1 Ma, and reached the acme of the first uplift phase from 10.9 to 7.5 Ma [1]. Micropaleontological studies of the deep sea cores from the Arabian Sea also indicated that the monsoonal upwelling caused by the uplift of the Himalayas was initiated from 10-9 Ma [10].

In the Churia (Siwalik) Group, the appearance of "Pepper-and Salt Sandstone" containing such minerals as kyanite and garnet from the Himalayan Gneiss, from the upper member of the Arung Khola Formation (10.5 Ma) probably indicates the uplifting of the High Himalayas [20; 23]. From the same time, the periodic flood deposits of meandering rivers dominate the Siwalik sedimentary sequence [8]. All this sedimentological evidence suggests that the Himalayas become elevated high enough to develop the monsoon system. During the deposition of the lower member of Binai Khola Formation (7.0 to 9.5 Ma), the depositional center shifted from the north to the south by uplifting along the Main Boundary Thrust (MBT). Consequently the dominant river system of the Siwaliks changed to braided rivers [8]. The Chitwan Formation (2.5 to 1.3 Ma) is mainly composed of coarser fan-system deposits and covers the fossiliferous beds. During the deposition of the Deorali Formation (after 1.3 Ma), the Churia sedimentary basins of this area were highly deformed and separated by the Central Churia Thrust (CCT; [20]). This is correlated to the second uplift phase of the Himalayas of Amano and Taira [1].

The change of vegetation caused by the global environmental change is revealed by the carbon isotope data. The δ^{13}C composition of pedogenic and organic carbon shows shifting from the lower to higher value at 7.7 Ma in Pakistan and at 7.0 Ma in the Surai Khola section, West Nepal [17; 18]. The tendency is recognized in our area, only the timing is slightly later than that in western areas (fig.3). The shifting means the displacement of C3 vegetation by C4 one. Quade et al. [18] insist that the floral change was probably continent wide and all the flood plains of major rivers were dominated by monsoonal grasslands by the beginning of the Pliocene.

FRESHWATER MOLLUSCS SUGGEST THE HISTORY OF THE HIMALAYAN UPHEAVAL

As reviewed above, the Himalayan upheaval and the consequential change of environment took place from ca.10 Ma, and at ca. 7 Ma with the onset of monsoonal conditions and vegetational change. The temporal and spatial distribution of the Siwalik

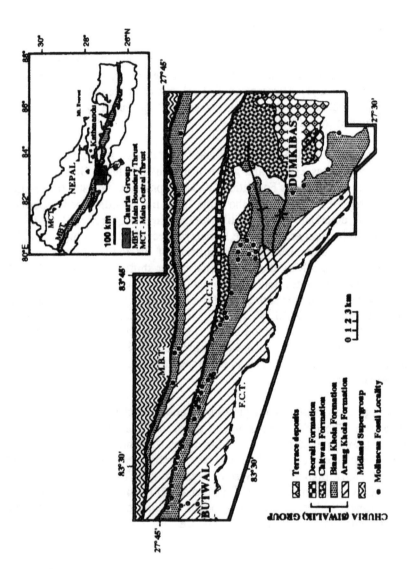

Figure 1. Index map and outline of geology

freshwater molluscs can be explained well in this framework.

The first occurrence of the freshwater molluscs in the Churia Group coincident with the acme of the first uplift phase of the Himalayas, is well indicated by the development of a meandering river system assumed from sedimentological data. The oxbow lakes and creeks in the back marsh of the rivers provided preferred habitats for freshwater molluscs. The taxa in the Indian subcontinent such as *Lamelllidens* and *Parreysia* invaded this habitat-set.

The main molluscan faunal change at about 7 Ma is also coincident with the change of vegetation and the development of the monsoon. The most probable cause of this faunal change is the migration of animals. According to Damme [6], the small-sized molluscs such as *Bithynia* and *Pisidium* can be carried by birds and other animals to distant places without any hydrographic connection. He also pointed out that freshwater pulmonates may by carried more easily by animals because of their adhesive egg clusters.

As for the recent climatic system, the establishment of the monsoonal condition means a climatic zonation similar to the recent one formed over the Eurasian Continent. This caused the differenciation of land environments, and consequently urged land animals to migrate or defuse to suitable places. The major faunal turnover in Pakistan marked by the beginning of the "*Hippalion* s.l." Intervalzone [3] is about 10.5 Ma on the recent geomagnetic timescale [5], which suggests the frequent immigration of the new fauna from Central Asia and Europe. After the following Intervalzone which is marked by the appearance of *Selenopoltax lydekkeri* (ca.8 Ma), many taxa including semi-aquatic mammals are derived from an African source [3]. The migration seems to relate strongly to the vegetational change in the flood plains from forests to grasslands.

In Nepal, unfortunately we have little information about the mammalian fauna. The tropical evergreen forest remained up to 9.5 Ma and thereafter semi-deciduous and dry deciduous forest became predominant [9; 18]. Consequently, the grassland expanded onto the flood plains by ca. 7 Ma as already mentioned. Corresponding to these vegetational changes, the mammalian migration must have taken place during the deposition of the Churia Group. Especially after 7 Ma, many kinds of mammals may have migrated to the open field along the rivers.

In addition to the role of land animals as a dispersal agent, the migratory birds are also important in this sense. Fossil evidence of birds has never been found from the Churia Group, much less any expectation of evidence of migration. The forming of climatic zonation and seasonal monsoon, however, provide enough to trigger the seasonal migration of birds, resulting in the introduction of palearctic or holoarctic genera such as *Pisidium, Bithynia*, etc.

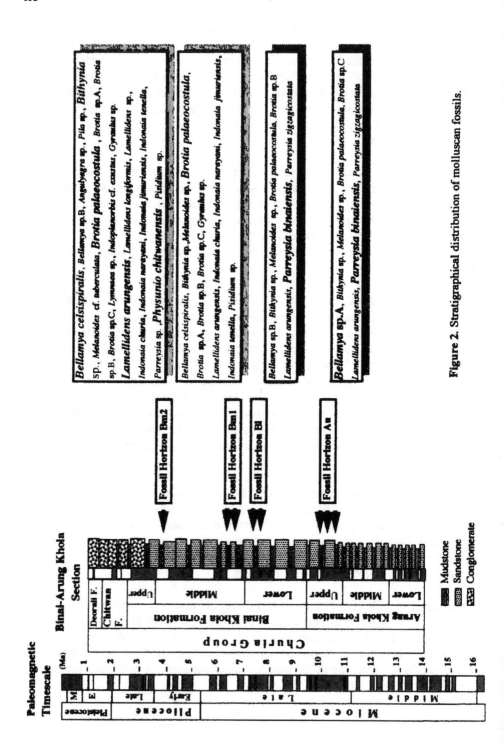

Figure 2. Stratigraphical distribution of molluscan fossils.

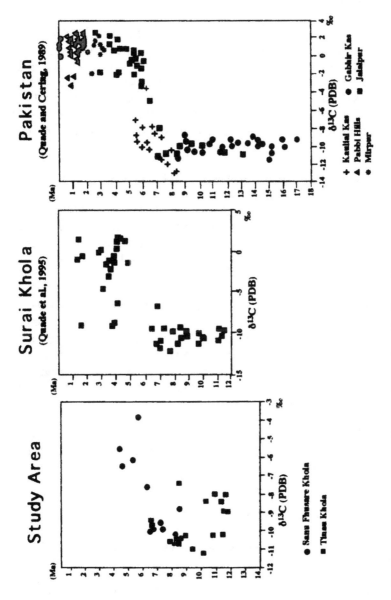

Figure 3. d13C composition of soil carbonate from the Tinau Khola and other areas.

CONCLUDING REMARKS

In this paper we have tried to show that the mode of occurrence of freshwater molluscan fossils from the Churia Group reflects the history of the Himalayan upheaval and its related environmental changes.

The first occurrence of molluscan fossils at about 10.5 Ma coincides with the onset of the monsoonal condition in the South Asia which was probably caused by the first phase of the Himalayan upheaval. Floods caused by the seasonal heavy rain may have changed the meandering river course and formed preferable environments for molluscs such as oxbow lakes, ponds and creeks. The faunal enrichment after 7 Ma may also have related to the animal migration caused by the climatic zonation. The marked climatic zonation can be also considered to have occurred as a result of the Himalayan upheaval.

According to the sedimentological results, sand-bed braided rivers became predominant after 7 Ma and changed to gravel bed-rivers of fans at about 3 Ma. For most freshwater molluscs, braided rivers are found unfavorable due to their unstable condition. Fossil localites, however, are found the most frequently at the Bmu Horizon (3.5 Ma). We do not have a clear answer as to why the horizon contains such numerous fossil localities. More detailed examination from the sedimentological and ecological viewpoint is required.

The process of dispersal of freshwater molluscs must also be examined closely. Though the faunal enrichment after 7 Ma can be explained by the introduction of the small-sized molluscs with active dispersal ability, the dispersal of large-sized bivalves, Unionidae, depends on the fishes for transportation, since they are parasitic to them in their larval stage. Therefore it is important to examine the hydrographic connection and fish distribution pattern for understanding the process of sequential faunal change of freshwater mollusc populations.

Although many problems still remain, as mentioned above, freshwater molluscs can be useful as an indicator of the prevalent environment because of their restricted habitat and mode of dispersal.

REFERENCES

1. K. Amano and A. Taira. Two-phase uplift of Higher Himalayas since 17 Ma. *Geology* 20, 391-394, (1992).
2. N. Annandale. Indian Fossil Viviparae. *Rec. Geol. Surv. India* 50, 362-367, pl.11, (1921).
3. J.C. Barry, E.H. Lindsey and L.L. Jacobs. A biostratigraphic zonation of the middle and upper Siwaliks of the Potwar Plateau of northern Pakistan. *Palaeogeography, Palaeoclimatology, Palaeoecology* 37, 95-130, (1982).
4. D.B. Bhatia and A.K. Mathur. Some Upper Siwalik and Late Pleistocene Molluscs from Panjab. *Himalayan Geology* 3, 24-58, (1973).

5. S.C. Cande and D.V. Kent. Revised calibration of the geomagnetic polarity timescale for the Late Cretaceous and Cenozoic. *Journal of Geophysical Research* 100: B4, 6093-6095, (1995).
6. D.van Damme. The freshwater mollusca of northern Africa. Developments in Hydrobiology, 25. Dr. W. Junk Publishers, Dordrecht, 164p., (1984).
7. D. Gurung, K. Takayasu and K. Matsuoka. Freshwater Molluscan Fauna of the Miocene-Pliocene Churia (Siwalik) Group of Nepal and their implication. *Proc. 30th Int'l. Geol. Congr.*, 12.
8. K. Hisatomi and S. Tanaka. Climatic and environmental changes at 9 and 7.5 Ma in Churia (Siwalik) Group, West Central Nepal. *Himalayan Geology* 15, 168-180. (1994).
9. M. Konomatsu and N. Awasthi. Flora of the Churia Group, Central Nepal. *Abstract Symp. Himalayan Geology, Shimane* 1992, Japan, 24, (1992).
10. D. Kroon, T. Steen and S.R. Troelstra. Onset of monsoonal related upwelling in the western Arabian Sea as revealed by planktonic foraminifers. In Prell, W.L., Niitsuma, N., et al., Proc. ODP, Sci. Results, 117: College Station, TX (Ocean Drilling Program), 257-263, (1991).
11. B. Prashad. On fossil ampullariid from Poon, Kashmir. *Rec. Geol. Surv. India* 56, 210-212, (1925).
12. B. Prashad. On a collection of land and freshwater fossil molluscs from the Karewas of Kashmir. *Rec. Geol. Surv. India* 56, 356-361, pl.29, (1925).
13. B. Prashad. On some fossil Indian Unionidae, *Rec. Geol. Surv. India* 60, 308-312, (1927).
14. B. Prashad. Recent and fossil Viviparidae-a sudy in distribution, evolution and paleogeography. *Mem. Ind. Mus.* 8, 153-152, (1928).
15. B. Prashad. On some undescribed freshwater molluscs from various parts of India and Burma. *Rec. Deol. Surv. India* 63, 428-433, (1930).
16. B. Prashad. Some freshwater and land fossil molluscs from near Ghorband, Afghanistan. *Rec. Geol. Surv. India* 72, 125-129, (1937).
17. J. Quade and T.E. Cerling. Expansion of C4 grasses in the late Miocene of northern Pakistan: Evidence from paleosols. *Palaeogeography, Palaeoclimatology, Palaeoecology* 115, 91-116, (1995).
18. J. Quade, J.M.L. Cater, T.P. Ojha, J. Adam and T.M. Harrison. Late Miocene environmental change in Nepal and the northern Indian subcontinent: Stable isotopic evidence from paleosols. *Geol. Soc. Amer. Bulletin* 107: 12, 1381-1397, (1995).
19. K. Takayuasu, D. Gurung and K. Matsuoka. Some new species of freshwater bivalves from the Mio-Pliocene Churia Group, west-central Nepal. *Trans. Proc. Palaeont. Soc. Japan, N.S.* 179, 157-168, (1995).
20. T. Tokuoka, K. Takayasu, M. Yoshida and K. Hisatomi. The Churia (Siwalik) Group of the Arung Khola Area, West Central Nepal. *Mem. Fac. Sci. Shimane Univ.* 20, 135-210, (1986).
21. T. Tokuoka, S. Takeda, M. Yoshida and B.N. Upreti. The Churia (Siwalik) Group in the western part of the Arung Khola area, west Central Nepal. *Mem. Fac. Sci., Shimmane Univ.* 22, 131-140, (1988).
22. T. Tokuoka, K. Takayasu, K. Hisatomi, H. Yamasaki, S. Tanaka, M. Konomatsu, R.B. Sah and S.M. Ray. Stratigraphy and geologic structures of the Churia (Siwalik) Group in the Tinau Khola-Binai Khola Area, West Central Nepal. *Mem. Fac. Sci., Shimane Univ.* 24, 71-88, (1990).
23. T. Tokuoka, K. Takayasu, K. Hisatomi, S. Tanaka, H. Yamasaki and M. Konomatsu. Tkhe Ckhuria (Siwalik) Group in West Central Nepal. *Himalayan Geology* 15, 23-35, (1994).
24. H.E. Vokes. Unionidae of the Siwalik Series. Mem. Connecticut Acad. 9, 37-48, (1935).
25. H.E. Vokes. Siwalik Unionidae from the collection of the Second Yale North India Expedition. *Quart. Jour. Geol. Min. Met. Soc., Indiana* 8, 133-144, (1936).
26. R.M. West, J. Jr. Munthe, J.R. Lukacs and T.B. Shrestha. Fossil mollusca from the Siwaliks of Eastern Nepal. *Current Science* 44, 497-498, (1975).

Proc. 30th Int'l. Geol. Congr., Vol. 12, pp. 136-146
Jin and Dineley (Eds)
© VSP 1997

The Change of Late Cenozoic Marine Macrofauna in the Eastern Margin of the Sea of Japan

IWAO KOBAYASHI[1], TETSURO UEDA[1], HIDEAKI NAGAMORI[2], TATSUYA SAKUMOTO[2], HIDEO YABE[2], MAKOTO MIYAWAKI[2], and MASAKAZU HAYASHI[2]

[1] Department of Geology, Faculty of Science, Niigata University, Nino-cho,Ikarashi, Niigata 95021, Japan
[2] Graduate School of Science and Technology, Niigata University, Nino-cho, Ikarashi, Niigata 950-21, Japan

Abstract

The Sea of Japan has existed since Early Miocene time and the environmental conditions have changed during the Neogene and Quaternary geographically and oceanographically. The history of the Sea of Japan is generally divided into six stages, 1) the age of continental margin (Paleogene to Early Miocene), 2) the age of inception of a marginal sea and great transgression (late Early Miocene to Early/Middle Miocene), 3) the age of deepening sea (Middle Miocene), 4) the age of deep sea (Middle to Late Miocene), 5) the age of formation of present marginal sea (late Late Miocene/Early Pliocene) and 6) the age of present marginal sea (Late Pliocene to Recent). In the second stage the coastal shallow sea was mainly covered by tropical and subtropical elements of marine macrofaunas. In the fourth stage marine faunas and floras changed mainly to temperate elements of Pacific Ocean with rare subtropical elements. Between the fourth and fifth stages the shift of species in various taxa was well recognized. In the fifth and sixth stages the migration of marine organisms was actively carried out from the southern and northern seas. The change of marine macrofaunas were strongly effected by the exchange of warm and cold sea currents, and the appearance and disappearance of land-bridges and straits with the sea level changes caused by climatic and crustal movements.

Keywords: marine macrofauna, Late Cenozoic, the Sea of Japan, mollusc, Decapod, Elasmobranch, marine mammal

INTRODUCTION

Very thick Tertiary and Quaternary strata including fossiliferous beds are widely distributed in the Sea of Japan and its surrounding land area (Figure 1). The geological history of the Sea of Japan since Early Miocene time has been described by many previous Japanese geologists. Palaeogeographical maps of several stages in the Neogene had been drawn by some of them [14]. Late Cenozoic marine fossils are represented by foraminifers, molluscs, crabs, fishes, mammals etc. They offer important geological, geographical and palaeontological informations related to the history of a marginal sea.

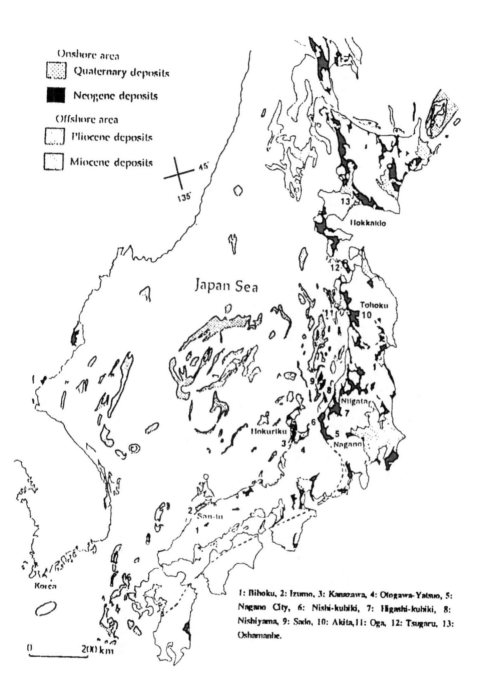

Figure 1. Distribution of Late Cenozoic strata

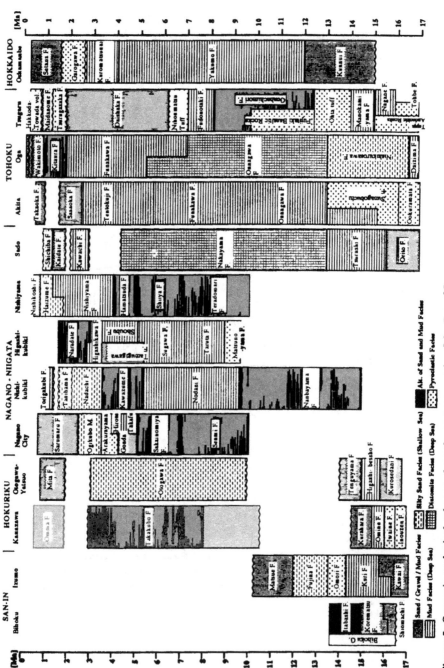

Figure 2. Cenozoic geological columns along the eastern margin of the Sea of Japan

OUTLINE OF GEOLOGICAL EVENTS

Figure 2 shows representative geologic columns of Neogene to Early Pleistocene deposits along the coast of the Sea of Japan. They indicate many different environments, from terrestrial to deep sea.

Important palaeogeographical events in the Sea of Japan [13] (Table 1), were 1) the inception of a marginal sea (late Early Miocene), 2) the appearance of an archipelago washed with the inflow of strong warm currents (Early/Middle Miocene), 3) the widening and deepening of the sea, the change to Pacific temperate sea and the beginning of a closed marginal sea (Middle Miocene), 4) The inferred active stage of the Sea of Japan (Late Miocene), 5) the beginning of the present marginal sea continuing to the recent sea (late Late Miocene/Early Pliocene), 6) the marginal sea effected by eustatic and crustal movements. Figure 3 shows palaeogeographical sketch of maps of the three stages. In this article, six stages are discriminated according to important historical events. The marginal sea (I) during Stage 1 to 4 (Table 1) was widely continued to the Pacific Ocean, especially in the northern part (Figure 3). On the other hand, the sea (II) during Stage 5 to 6 was surrounded by the uplifting lands (e.g. Honshu Island) with narrower straits in the northern and southern parts similar to the present sea.

Stage 1: the age of continental margin (Paleogene to Early Miocene)
During Paleogene and early to middle Early Miocene time, there was no marginal sea. The stage locally had strong volcanic activities, characterized by as green volcanic tuff. There were several lakes in which many kinds of fossils such as plants, insects and fishes were deposited.

Stage 2: the age of inception of a marginal sea and great transgression (late Early to Early/Middle Miocene)
During late Early to Early/Middle Miocene time, the great transgression after active crustal movements accompanied with depressions and the inception of a sea occurred along the continental margin [3,14]. Warm currents of Pacific Ocean went up to Hokkaido, where the north latitude is about 44 degrees with tropical and subtropical faunas. An archipelago existed in the shallow sea.

Stage 3: the age of deepening sea (early Middle Miocene)
During early Middle Miocene time, the sea became deeper and widened with basic submarine volcanic activity, and changed to a temperate sea.

Stage 4: the age of deep sea (Middle to Late Miocene)
During Middle to Late Miocene time, deep sea basins were formed in the sea which widely opened to North-West Pacific Ocean. In Niigata, the sea was more than 2,000m deep and a large-scale submarine fan formed. The sea was temperate to boreal with abundant marine plankton such as diatoms. In the middle of this stage, 10Ma, warm currents temporarily flowed into the marginal sea. At the end of this stage, the major uplifting of the island arc and the growth of backbone ranges occurred. As the result of these crustal movements, the present marginal sea was formed.

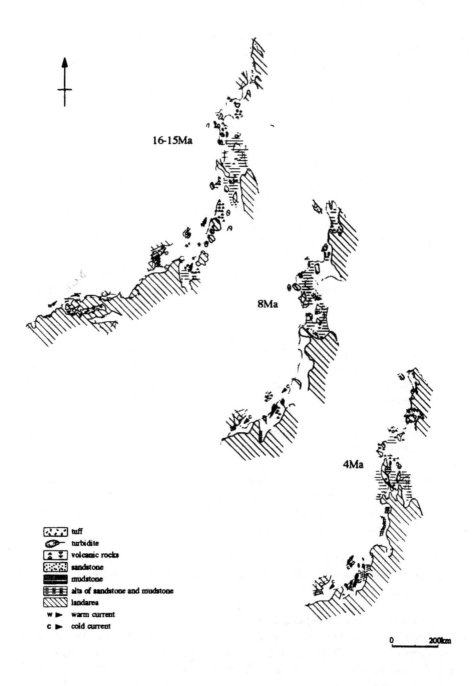

Figure 3. Palaeogeographical maps along the eastern margin of the Sea of Japan

Table 1. Geologic events and stage divisions in the Sea of Japan

Period	Epoch	Marginal Sea	Age	Events	No.
Quaternary / Pliocene		MARGINAL SEA II	The age of PRESENT MARGINAL SEA	Crustal and climatic sea level change / Inflow of warm and cold currents	6
				Formation of submarine fans / Uplifting and subsiding sea bottom / Uplifting of Japanese Island	5
Miocene	Late / Middle	MARGINAL SEA I	The age of DEEP SEA	Growth of backbone ranges / Sea widely opening to the north-west Pacific Ocean / Temperate to boreal sea / Deposition of marine plankton / Sea depth more than 2,000m / Formation of great submarine fans	4
			The age of DEEPENING SEA	Change to temperate sea / Basic submarine volcanic activity	3
	Early		The age of GREAT TRANSGRESSION	Tropical-subtropical sea	2
			The age of CONTINENTAL MARGIN	— Inception of Marginal Sea — / Strong volcanic activity on land / Lake in existence	1

Table 2. Macrofaunal succession in the Sea of Japan

Period	Epoch	Marginal Sea	Fauna	No.
Quaternary / Pliocene		MARGINAL SEA II	Temperate - boreal fauna (Omma - Manganji fauna)with warm current fauna	6
				5
Miocene	Late	MARGINAL SEA I	Cool - Temperate fauna (Shiobara fauna)	4
	Middle		Temperate fauna (Shiobara fauna)	3
			Tropical and subtropical fauna (Kadonosawa fauna)	2
	Early		Fresh water fauna	1

Stage 5 and Stage 6: the age of present marginal sea (late Late Miocene to Recent)
From late Late Miocene to Recent time, uplifting and subsiding sea floor with troughs spread, and many small-scale submarine fans were formed along the coastal area of the northern Honshu. Alternating inflows of warm and cold currents accompanied climatic sea level changes.

REPRESENTATIVE MARINE MACROFAUNAS

Molluscan fossils: Figure 4 shows main molluscan faunas in the Sea of Japan during Neogene [1, 2, 12, 15, 17]. The Early and Middle Miocene Kadonosawa Fauna was representative of tropical and subtropical elements widely distributed up to Hokkaido. The Middle and Late Miocene Shiobara Fauna adapted to a temperate sea in West Pacific. The Pliocene and Early Pleistocene Omma - Manganji Fauna including temperate and cold elements was characteristic of the Sea of Japan.

Decapod crustacean fossils: The occurrence of Miocene decapod fauna is stratigraphically divided into two horizons, early Middle Miocene [8, 9, 10], and middle Middle Miocene [16]. The former is mainly composed of tropical elements, *Thalassina anomala* and *Ozius collinzi* [8, 10], and subtropical to temperate elements [9]. The latter is mainly composed of a cold water element, *Cancer (Metacarcinus) izumoensis*, and subtropical to temperate elements [16]. A Pliocene fauna is not well-preserved.

Elasmobranchian fossils: *Isurus desori, Carcharodon carcharias, Carcharocles megalodon* and *Carcharhinus* spp. [7, 18, 19] are dominant during Neogene. Their occurrences are divided into three horizons, namely Early to Middle Miocene, middle Middle Miocene to Late Miocene, and Late Pliocene to Pleistocene ages. There was an important change between the latter two ages.

Marine mammal fossil: The important taxa are Desmostylia, Ziphiidae, Delphinidae and Balaenopteridae [11]. *Desmostylus* and *Paleoparadoxia* [6] were widely distributed along the northern Pacific region during Early to Middle Miocene. They lived on the shores of the archipelago. Cetacean remains were sometimes found in early Middle Miocene, Late Miocene and Pliocene strata.

CHANGE OF MARINE MACROFAUNA

Stage 1
The climate was cool-temperate and the land was covered by deciduous-broad leaf and pine trees, named as the Aniai-type flora [5], such as *Alnus, Betula, Carpinus* and *Acer*, accompanied with insects, Pentatomidae [4]. Fresh-water fishes such as carp and planktonic diatoms such as *Melosira* lived in the ancient lake.

Stage 2
During this stage, the archipelago was engulfed by strong warm currents. Many tropical and subtropical species went up northwards with the current. Tropical marine faunas reached to the central area of Honshu Islands and subtropical marine faunas went up to

Hokkaido. The Kadonosawa Fauna was composed of tropical and subtropical shallow marine, and mangrove-swamp elements such as *Geloina* and *Telescopium* . Decapods are tropical elements, e.g.*Thalassina anomala, Ozius collinzi*, and several species of subtropical and temperate elements [8, 9, 10]. These species seem to have lived in intertidal to lower sublittoral zones. Elasmobranchs, *Isurus desori, Carcharodon carcharias, Carcharocles megalodon* and *Carcharhinus* spp., *Synodontaspis, Isurus* and *Carcharocles* [7, 18, 19] and marine mammals, Ziphiidae, Delphinidae and Balaenopteridae thrived in the open sea. Marine-coast dwelling large mammals, *Desmostylus* and *Paleoparadoxia* also survived.

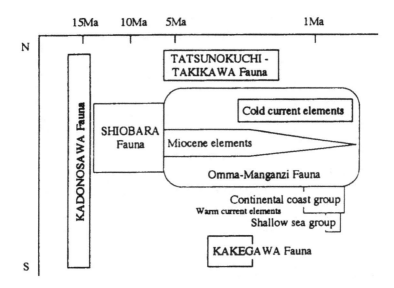

Figure 4. Miocene to Pleistocene molluscan fauna in the Sea of Japan

Stage 3
Deepening and widening of the marginal sea occurred after the great transgression. Marine climate was transitional, tropical to temperate. The molluscan fauna changed to the Shiobara Fauna which adapted to a temperate climate. Deep sea molluscs such as *Palliolum* and *Solemya* appeared in the deep sea. Decapods, the *Carcinoplax antiqua* Assemblage lived in subtropical to temperate seas and in the sublittoral zone. The representatives of a cold water element, *Cancer (Metacarcinus) izumoensis*, and subtropical to temperate elements, such as *Itoigawaia minoensis, Scylla ozawai* and *Carcinoplax antiqua* [16] are found. Reconstructing the deep sea biota, marine phyto-planktons were very flourishing. Marine fishes and mammals such as Ziphiidae, Cetotheriidae and *Berardius* were abundant in the open sea. *Desmostylus* and *Allodesmus* also existed along the coasts. Large sharks such as *Carcharocles megalodon* also were abundant in the sea [7, 19].

Stage 4
The climate of the marginal sea changed to cool-temperate. In the shallow sea, the Shiobara Fauna was temperate dwellers. Decapods was represented by the *Carcinoplax*

antiqua Assemblage which is thought to have lived in a sublittoral zone. *Mursia takahashii* was a cold element. On the other hand, in the deep and open sea, marine phyto- and zoo-plankton were very flourishing, but a deep sea benthic fauna was very limited in composition. Arenaceous foraminifers, *Cyclammina* and *Haplophragmoides* lived on the deep sea bottom under anaerobic condition, and sometimes trace fossils like chondrites are found from deep sea deposits. One kind of cephalopod came to the deep sea, riding on warm currents.

Table 3. Marine fauna of Stage 6 (Late Pliocene - Early Pleistocene)

TAXA	TIDAL ZONE	SHALLOW SEA	DEEP SEA
elasmobranchs		*Carcharodon carcharias, Isurus oxyrinchus, Carcharhinus*	
marine mammals		*Hydrodamalis, Eumetopias*	*Mesoplodon, Orcinus, Balaenopteridae*
molluscs. decapods. sea urchins. brachiopods, bryozoa	Continental Coast Moll.Fauna	Omma-Manganji Moll. Fauna	*Palliolum peckhami*
zooplankton		Planktonic foraminifer	
phytoplankton		Diatomacea	

Stage 5

During late Late Miocene to Early Pliocene time, the marine climate changed to temperate. The marginal sea similar to the recent one in shape was formed with rising sea floors and subsiding troughs. Marine faunas of warm temperate and shallow sea taxa were dominant. In the shallow sea, the Omma Manganji Fauna which adapted to cool-temperate conditions spread widely in the Sea of Japan. The Tatsunokuchi Fauna appeared in the northern sea. Marine mammals such as *Protodobenus*, Otariidae, *Sirenia*, also were seen with many kinds of Cetacea. *Carcharodon carcharias* and *Parotodus benedeni* now made an appearance [7, 19].

Stage 6

The sea continued from the previous age with warm temperate and cold faunas of littoral to sublittoral elements and deep sea benthonic faunas (Table 3). The alternating climate changes, warm and cool, weakly and strongly occurred with sea level changes and exchanges of warm and cold currents. Marine warm temperate and shallow sea faunas flourished. At 3Ma, planktonic warm species came from the southern sea through the Tsushima-Korea straits. The Omma-Manganji Fauna had survived since Early Pliocene time, and the continental coast molluscan fauna was added especially from Early Pleistocene. Marine mammals such as *Sirenia* and Odobenida migrated from the cold sea to the southern sea. *Mesoplodon, Orcinus, Carcharodon, Isurus* and *Carcharhinus* lived in abundance [7, 19].

CONCLUSION

During Neogene and Early Pleistocene time, the change of faunas or assemblages in the Sea of Japan was profoundly effected by the change of sea currents, warm and cold, and the palaeogeographical consequences of sea-level changes. In Stage 2, tropical and subtropical faunas migrated to the northern area as transgression occurred. In Stage 3, the tropical and subtropical marine fauna disappeared and in contrast, temperate faunas appeared due to the change of sea currents. In Stage 4, the Central and North Pacific shallow sea biota which adapted to cool-temperate, the so-called Middle Miocene type biota may have been established in the North Pacific. In Stage 5 and 6, advanced Miocene species appeared in the various taxa. The change from Miocene type to Pliocene-Quaternary type occurred in various taxa. The Pliocene Omma-Manganji Fauna included many descendants of Middle Miocene types, for example the Shiobara Fauna. Next, alternations of marine macrofaunas were clearly related to the exchange of sea currents or water masses, which was caused by the opening and closing of straits and the global change of ocean climates. The species-increase of some taxa happened at Stage 2 and the end Stage 4 and 6 respectively. The arrival of temperate species is very significant and the species are thought to be ancestors of recent species. During Early Pleistocene, the continental coast fauna spread to the northern region. Most of them are intimately related to recent species.

REFERENCES

1. K. Chinzei. Neogene molluscan faunas in the Japanese Islands- an ecologic and zoogeographic synthesis, *The Veliger* **26**, 155-170 (1978).
2. K. Chinzei. Marine biogeography in northern Japan during the early Middle Miocene as viewed from benthic molluscs. *Palaeont. Soc. Jap. Spec. Papers* **29**, 161-171 (1986).
3. Y. Fujita. The formation of Japanese Islands, - Circum Pacific tectonics, Tsukiji-shokan, Tokyo, 259p. (1990) .(in Japanese) 4. I. Fujiyama. Early Miocene insect fauna of Seki, Sado Island, Japan, with note on the occurrence of Cenozoic fossil insects from Sado to San-in district. *Mem. natn. Sci. Mus.* **18**, 35-59 (1985).(in Japanese with English abstract)
5. K. Hujioka. The Aniai flora of Akita Prefecture, and the Aniai-type floras in Honshu, Japan. *Jour. Min. Coll. Akita Univ.* [A] **3**, 1-105 (1964).
6. N. Inuzuka. Studies and problems on the Order Desmostylia. *In*: N. Inuzuka, K. Takayasu, K. Chinzei and K. Yoshida (Eds.), *Desmostylians and their Paleonvironment. Assoc. Geol. Colabor. Japan, Monogr.* **28**, 1-12 (1984).(in Japanese with English abstract)
7. H. Karasawa. Late Cenozoic elasmobranchs from the Hokuriku district, central Japan. *Sci. Rep. Kanazawa Univ.* **34**, 1-57 (1989).
8. H. Karasawa. The crab *Ozius collinsi* sp. nov. (Xanthoidea: Decapoda: Crustacea) from the Miocene Katsuta Group, southwest Japan, *Tertiary Res.* **14**, 19-24 (1992) .
9. H. Karasawa. Cenozoic decapod crustacea from southwest Japan. *Bull. Mizunami. Fossil. Mus.* **20**, 1-92 (1993) .

10. H. Karasawa and I. Nishikawa. *Thalassina anomala* (Herbst,1804) (Thalassinidea; Decapoda) from the Miocene Bihoku Group, southwest Japan. *Trans. Proc. Palaeont. Soc. Japan, N.S.* **162**, 852-860 (1991).
11. M. Kimura. Stratigraphy and inhabited environments of the Cetacea in Japan, *Mem. Geol. Soc. Japan* **37**, 175-187 (1992). (in Japanese with English abstract)
12. I. Kobayashi. Character and development of the Omma-Manganji fauna in the Niigata oil-field, Central Japan. *Palaeont.Soc. Jap. Spec. Papers* **29**, 245-255 (1986).
13. I. Kobayashi and M. Tateishi. Neogene stratigraphy and paleogeography in the Niigata region, Central Japan. *Mem. Geol. Soc. Japan* **37**, 53-70 (1992).(in Japanese with English abstract)
14. M. Minato, M. Gorai and M. Hunahashi (Eds). *The Geologic Development of the Japanese Islands*, Tsukiji Shokan, Tokyo, 442p.(1965)
15. K. Ogasawara. Notes on origin and migration of the Omma-Manganzian fauna, Japan. *Palaeont. Soc. Jap. Spec. Papers* **29**, 227-244 (1986).
16. T. Sakumoto, H. Karasawa and K. Takayasu. Decapod crustaceans from the Miocene Izumo Group, southwest Japan. *Bull. Mizunami. Fossil. Mus.* **19**, 441-453 (1992).(in Japanese with English abstract)
17. T. Ueda, K. Takayasu and I. Kobayashi. Cenozoic shallow-marine faunas of the coastal area along the Sea of Japan. *Saito Ho-on Kai Spec. Pub.* **3**, 515-528 (1991).
18. T. Uyeno and H. Uematsu. Middle Miocene elasmobranchs from Sunagawa, Yamagata Prefecture. *Mem. Natn. Sci. Mus.* **17**, 35-38 (1984). (in Japanese)
19. H. Yabe and I. Kobayashi. Atlas of the fossils from Niigata Prefecture: 5 Cenozoic elasmobranchs (sharks, rays and skates). *Jour. Geol. Educ. Res. Niigata* **28**, 33-44 (1994).(in Japanese)

Proc. 30th Int'l. Geol. Congr., Vol. 12, pp. 147-160
Jin and Dineley (Eds)
© VSP 1997

Palaeosynecological Development of Upper Anisian (Middle Triassic) Communities from Qingyan, Guizhou Province, China - a Preliminary Summary

FRANK STILLER

Geologisch-Paläontologisches Institut, Corrensstraße 24, D-48149 Münster, Germany

Abstract

A summary of the palaeosynecological development during the Upper Anisian of Qingyan (Guizhou Province, SW-China) is based on quantitative analyses of different successive fossil associations. A facies of deeper water with only few pelagic organisms developed into a shallow marine facies characterized by very rich, highly diverse benthic communities. Initially, several brachiopod-dominated communities of relatively low diversity were established. Later, highly diverse communities dominated by small bivalves and gastropods began to flourish. These very probably were bound to meadows of macroalgae. After some fluctuations and another alternation of these two types of communities, the facies developed in response to deeper-water conditions with only few pelagic organisms.

Keywords: palaeosynecology, Qingyan, Guizhou, China, Middle Triassic, Anisian

INTRODUCTION

Qingyan is a small town about 30 km S of Guiyang, the provincial capital of Guizhou Province in southwestern China (Fig. 1). The main study area marked on the map is situated NNE of Qingyan and covers a strip from the river to the foot of Shizishan (Mountain). It is equivalent to a section through the Upper Anisian which is called Yuqing Subformation in the local stratigraphy (Fig. 2). The main locality investigated is the highly fossiliferous section at Bangtoupo - wrongly also called "Leidapo" -, stratigraphically in the middle part of the Leidapo Member which is of early Late Anisian age. Here the rare opportunity exists to study a sequence of rich and highly diverse fossil associations in great detail and with high resolution. The Upper Anisian of Qingyan has a thickness of 396 m [2], the Bangtoupo section comprises about 50 m. At Bangtoupo the Anisian rocks are well-exposed, but in its vicinity the outcrop conditions are much poorer due to agricultural utilization of all available ground.

Previous Work

Qingyan and particularly Bangtoupo ("Leidapo") has been known for a long time as an important site of marine Middle Triassic fossils in SW-China. It was first mentioned by Koken [8] who described various fossils from this locality not collected by himself. Xu and Chen [17] published a revised list of fossils from Qingyan, but gave neither

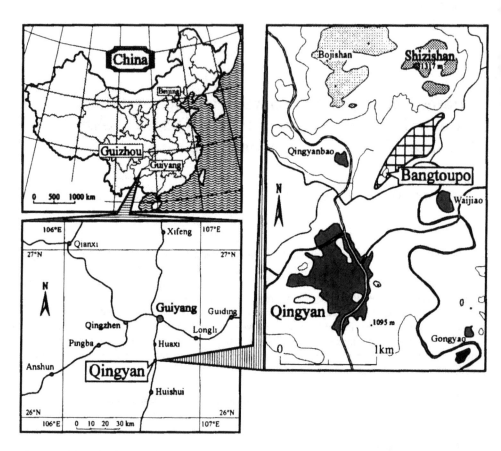

Figure 1. Locality map of Qingyan and its vicinity, Guizhou Province, SW-China. Vicinity-map of Qingyan with topography; main study area NNE of Qingyan with Bangtoupo chequered; villages and rivers/brooks in black.

figures nor sufficient descriptions of the fossils. During the sixties to the eighties, several fossil groups were studied separately and isolated fossils from Qingyan and "Leidapo" were described. The articles are mostly written in Chinese, not comprehensive, and are often scattered over extensive collective volumes [e.g. 3, 6 (parts), 9 (parts), 10 (parts), 18, 21-23]. In spite of the multitude of studies of the fossils from Qingyan and the importance of this site, the treatment has remained incomplete so far. A first quantitative summary of the palaeosynecology of the fauna from Bangtoupo was presented by Stiller [12].

The scope of the ongoing project is a comprehensive study of the Upper Anisian macrofauna from Qingyan, especially from the Bangtoupo section, including detailed quantitative reconstructions of the palaeosynecology of the different fossil associations and of their historical development. This paper sketches the palaeosynecological development; taxonomical studies of the fossils and detailed analyses of the palaeoecology of the different associations will be presented in the near future.

Middle Triassic	Ladinian	Guiyang Formation		Ganyintang Member	>362 m
				Shizishanjiao Member	59 m
	Anisian	Qingyan Formation	Yuqing Subformation	Yuqing Member	204 m
				Leidapo Member *Bangtoupo*	192 m
			Xiaoshan Subformation	Yingshangpo Member	176 m
				Mafengpo Member	128 m
				Xiaoshan Member	131 m

Figure 2. Stratigraphic table of the Middle Triassic in the Qingyan facies region. (Thicknesses of the different members after [2].) Upper Anisian set off; position of the Bangtoupo section marked

METHODS

Several successive bulk samples were collected along the well-exposed, highly fossiliferous Bangtoupo section. The fossil content of 23 bulk samples forms the basis of the statistical analyses. From the Upper Anisian strata beneath and above this section, only non-statistical samples could be obtained. Larger fossils were prepared and cleaned mechanically, then the remaining rock material was soaked in water; and after disintegration it was sieved in order to obtain the small fossils. All fossil remains larger than 1-2 mm were included in the study, microfossils were not investigated. After identification of the fossils the number of statistical specimens (n) was calculated for each taxon. In the case of bivalved organisms (bivalves, brachiopods) the number of isolated right or left or ventral or dorsal valves was added to the number of articulated specimens, depending on which one was higher. In the case of colonial organisms, each colony was counted as one individual. Several conspecific larger fragments were combined and counted as one valve or specimen respectively. The number of crinoids was calculated by taking the largest number of either holdfasts or stems, whereby 150 columnals were regarded as equivalent to the average length of stem. Due to the generally problematic identification of isolated skeletal elements of echinoids and the poor knowledge of Anisian echinoids, the echinoid remains were grouped into several morphospecies only. 35 spines or interambulacral plates were regarded to represent one individual. The number of spine or plate morphospecies was taken as the number of species, depending on which one was higher. The number of individuals was calculated correspondingly. In the case of small encrusting taxa like serpulids or bryozoans, the occurrence on each host was counted as one individual. For the calculation of the size-adjusted relative abundance of each taxon (na), the taxa were grouped according to their size and the number of statistical individuals (n) was multiplied with the size factor (very very small (serpulids) = 0.2-0.3; very small = 1; small = 2; medium = 4; large = 8;

very large = 16). This way the biovolume as an approximation of biomass was estimated.

FACIES AND TAPHONOMY AT BANGTOUPO

The Bangtoupo section comprises a highly fossiliferous sequence of thinly-bedded, more or less marly and mostly slightly silty mudstones with thin more calcareous intercalations (marls to marly limestones) deposited under shallow marine conditions.

The fossils from Bangtoupo generally are well- to very well-preserved retaining their calcareous shells. Only few gastropod taxa sometimes occur as steinkerns. This indicates that larger-scale diagenetic bias is very unlikely.

Epibenthic bivalves mostly are preserved disarticulated, endobenthic bivalves often articulated, just as are many brachiopods. Echinoderm taxa are mostly preserved in form of disarticulated skeletal elements, but there are also articulated segments of crinoid stems and of echinoid interambulacra. Even a complete echinoid corona with spines sticking to it was discovered. Most shells were embedded without damage, or at least almost so. After being finally buried, however, many shells were deformed or broken by compaction and/or tectonic pressure.

There has been no transport sorting according to size or form. This is indicated by the common occurrence of a very wide spectrum of sizes and shapes of fossils, the great taxonomic diversity, and the presence of juvenile to adult individuals of relatively abundant taxa. Also, both valves of disarticulated bivalved taxa were generally recorded without noticeable dominance of one type of valve. With the possible exception of some rounded or corroded corals, sponges, crinoid ossicles and calcareous algae, there are no preservational features suggesting transport. Generally no current-induced uniform orientation of skeletal elements and no physically concentrated skeletal remains occur in the section. The relative abundance of skeletal elements encrusted by epibionts is rather low, only the valves of larger bivalves sometimes are encrusted more heavily.

These taphonomic characteristics and the fine-grained nature of the sediment indicate a rather low-energy environment and exclude a significant lateral transport or in-situ reworking. Sporadic short-term high-energy events such as storms in a shallow marine environment caused turbulence without greater lateral transport of organisms and skeletal elements, disarticulation and damage of shells and other skeletal remains, and sometimes rapid burial (echinoid corona). The generally low importance of encrusting organisms shows that at least smaller skeletal elements did not rest on the seafloor for a longer time.

The benthic fossils thus are autochthonous in the sense that they were not exported from their own biotope before final burial; the associations therefore represent isotopic taphocoenoses. This allows detailed palaeosynecological reconstructions of the former communities and of their development, based on quantitative statistical analyses of the fossil content of bulk samples. As in previous attempts, such reconstructions rely on the hard parts mostly. Taphocoenoses are always incomplete and moreover somewhat time-averaged. Indirect indications and comparisons with recent conditions still permit further interpretations, which however always are more or less speculative.

The composition of the associations of benthic fossils preserved at Bangtoupo to a great extent mirrors initial conditions, they are not altered very much taphonomically, thus they are authentic relic associations. The term association is used in this sense for the associations from Bangtoupo. The term community is equivalent to biocoenosis and is used in this paper for reconstructed communities which also comprise organisms not preserved as fossils.

SEQUENCE OF FOSSIL ASSOCIATIONS AND FACIES DEVELOPMENT

During Middle Triassic times, a curved NE-SW-running belt of facies change existed in Guizhou, separating an extensive, more or less restricted carbonate platform in the NW from an open basin with mainly clastic sedimentation in the SE which was connected to Palaeotethys. The line of facies change was marked by a long, narrow belt of algal bioherms stretching along the edge of the carbonate platform. (Fig. 3) [4, 7, 11, 13, 14, 19, 20] The Upper Anisian sequence of Qingyan was deposited on the transitional slope from the carbonate platform to the basin.

In the lower part of the Leidapo Member, the Upper Anisian starts with thinly-bedded mudstones which ield few pelagic fossils (bivalves (*Posidonia, Daonella*), cephalopods) and/or ichnofossils from a few horizons only. From this deeper-water facies, a unit deposited under shallow marine conditions develops, particularly well-exposed at Bangtoupo. These thinly-bedded, more or less marly and mostly slightly silty mudstones with thin more calcareous intercalations (marls to marly limestones) contain the rich and highly diverse shallow marine fossil associations studied quantitatively. They represent mainly epibenthic, stable-soft-bottom associations very rich in mostly small taxa. Finally a transition takes place from the shallow marine conditions back to a facies of deeper water, represented by a very thick sequence mainly of thinly-bedded mudstones comprising the upper part of the Leidapo Member and the whole Yuqing Member. In the Yuqing Member especially, these mudstones can become marly in parts. This sequence again yields few mainly pelagic fossils (bivalves (*Posidonia, Daonella*), cephalopods) and/or ichnofossils at a few levels only. Because of the poor outcrop conditions it is not possible to study this transition in detail and quantitatively.

The Fossil Associations from Bangtoupo
The shallow marine associations from Bangtoupo are dominated by brachiopods, bivalves and gastropods. Brachiopods show a comparatively low diversity, but some of their species can be very rich in individuals. On the other hand, bivalves and gastropods often are highly diverse without a single species dominating noticeably. These three groups will be analysed separately below. Sponges, corals, annelids, scaphopods, bryozoans, crinoids, echinoids, ammonites, nautilids, and calcareous algae to variable

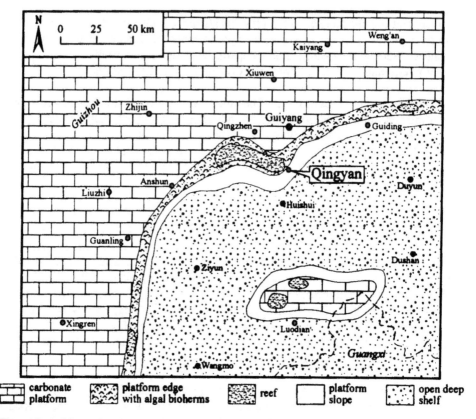

Figure 3. Anisian palaeogeography in central and southern Guizhou (after [11 and 7]).

degrees constitute only small parts of the fossil associations and will be dealt with as one unit for clarity. The microfauna consisting of foraminifers, ostracods, ophiuroid remains and others has not been studied in detail, as mentioned above. Fig. 4 for each sample/association gives the size-adjusted relative abundance of the four units brachiopods, bivalves, gastropods and "other taxa", the relative abundance (not size-adjusted) of brachiopods, bivalves and gastropods, and the species diversity.

The first sample ({1}) still shows clear influence of the facies of deeper water below. It represents a mixture of pelagic bivalves (*Posidonia*) and shallow marine benthic elements. Sample {2} already contains a clearly shallow marine benthic association without dominant species, consisting mainly of bivalves, brachiopods and to a smaller degree gastropods.

Initially, brachiopods clearly are the most important group, and several different brachiopod-dominated associations of comparatively low diversity develop. In these associations, brachiopods form far more than 50% of the association (na=61,8-77,5%). In the beginning, *Nudispiriferina minima* Yang et Xu is strongly dominating ({3}), then a transition to associations dominated by *Diholkorhynchia sinensis* (Koken) takes place ({5}, {6}).

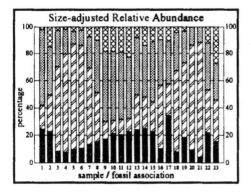

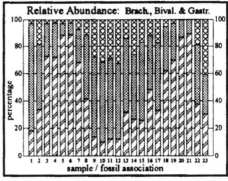

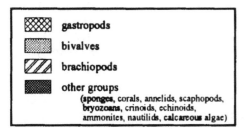

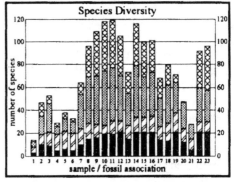

Figure 4. Size-adjusted relative abundance of brachiopods, bivalves, gastropods and "other taxa", relative abundance (not size-adjusted) of brachiopods, bivalves and gastropods, and species diversity for each sample/association.

With the association in sample {7}, the decline of the brachiopod-dominated associations commences. Brachiopods are still the most abundant group, but bivalves and gastropods become more important with regard both to relative abundance and to species diversity. As in the following associations, there is no single species dominating any more. In association {8}, diversity and relative abundance of bivalves and gastropods have increased further, and bivalves are the most abundant group now, followed by brachiopods.

This development leads to the highly diverse associations of mostly small benthic invertebrates in the middle part of the Bangtoupo section. These associations contain up to almost 120 species per sample and are dominated by bivalves and gastropods very clearly (sum of na=just under 70%) ({9}-{12}). Both groups show nearly the same number of species. The high species diversity of small gastropods is particularly striking. The importance of the brachiopods has diminished very much. No single species gains a prominent position in these associations. The associations of samples {9} to {12} all have to be assigned to this type, although they differ slightly in their individual composition. In these four associations the comparatively frequent occurrence of small calcareous algae (*Solenopora*) is remarkable.

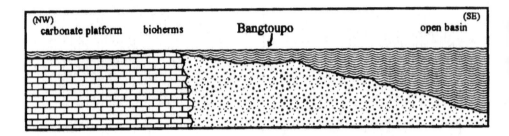

Figure 5. Palaeogeographic position of the protected shallow marine habitat of the communites from Bangtoupo in the upper part of the slope (not to scale).

An association with strongly reduced diversity compared to the previous units, in which the gastropods have lost a great part of their importance ({13}), is followed by more associations of high diversity without dominant species ({14}-{16}). Initially, bivalves are the most important group; gastropods form an important part of the associations, too. In association {14} small calcareous algae are unusually abundant again. In {16}, brachiopods replace bivalves as the most abundant group, but species diversity still remains high.

With slight variations, the general trend is towards a new phase of brachiopod-dominated associations of low diversity. In association {17}, bivalves once more are the most abundant group and small calcareous algae are relatively frequent again. The association of sample {18} still shows a relatively high diversity, but it is clearly dominated by brachiopods already. The most abundant brachiopod is *Mentzelia subspherica* Yang et Xu. In the following associations, brachiopods further gain importance, whereas the importance of bivalves and especially of gastropods just as overall species diversity decrease more and more. A remarkable phenomenon is the relatively great importance of corals (about 7%) in association {19}. Associations {20} and {21} are strongly brachiopod-dominated (na=76,7%, 84,3%), show a very low diversity, and are clearly dominated by *Diholkorhynchia sinensis* (Koken). In sample {21} its relative abundance even amounts to about 70% of the whole association.

Towards the top of the Bangtoupo section the compositional trend reverses. Bivalves and gastropods become important again, species diversity is high, and there is no dominant species any more. All in all, bivalves, gastropods and brachiopods are of similar importance. But the species number of these associations does not quite reach the diversity of samples {9}-{12}. Besides, brachiopods have a much greater and bivalves a smaller importance than in those samples.

In spite of the similarities seen in this superficial inspection, the different associations are each dominated or at least determined by different species. This applies to the various brachiopod-dominated associations as well as to the highly diverse bivalve-gastropod associations in different parts of the section.

Figure 6. Schematic palaeosynecological reconstruction: algal meadow community of high diversity.

PALAEOSYNECOLOGICAL INTERPRETATION

All the shallow marine benthic associations from Bangtoupo mainly consist of epibenthic organisms. Shallow burrowing infauna and semiinfauna form smaller portions. Nectic cephalopods are very rare. Most organisms are sessile. Mobile epifauna mostly consists of gastropods. With regard to trophic habits, suspension-feeders are the clearly dominant group. In addition, there are herbivores (gastropods), detritus-feeders, scavengers, and omnivores, few deposit-feeders (Nuculids), and few micro- and macrocarnivores. The associations are made up of organisms requiring soft substrates as well as others requiring firm or hard substrates. Cemented organisms could only make use of secondary hard substrates: they are attached to shells or skeletal remains of other organisms.

The composition of the highly diverse and rich fossil associations of Bangtoupo and the taphonomic situation there indicate a shallow marine environment of normal salinity and generally low energy, which was exposed only sporadically to short events of higher energy in the form of storms. These conditions are available in a protected shallow marine habitat in the upper part of the slope from the carbonate platform mentioned to the basin, perhaps on a terrace or more likely in a small basin in such a position (Fig. 5). The organisms lived in the photic zone above the storm wave base and below the fair-weather wave base. During Middle Triassic times, the climate in this was

Figure 7. Schematic palaeosynecological reconstruction: community of comparatively low diversity living on a seafloor poor in vegetation.

subtropical-arid to tropical-arid [15]. The ocean water was warm. The high percentage of suspension-feeders in the associations indicates clean and clear, gently moving shallow water rich in oxygen and nutrients. The sea bottom was fine-grained and stable. The rates of sedimentation were generally relatively low, presumably a steady slight influx of sediment particles prevailed.

The highly diverse associations of small invertebrates, especially bivalves and gastropods, most probably represent relics of communities which lived in meadows of macroalgae (Fig. 6). Algal meadows alter the environment in various ways and provide a highly differentiated and protected habitat rich in nutrition for a very diverse fauna of small invertebrates, corresponding to recent seagrass meadows [1]. Especially the high diversity of small gastropods indicates a habitat of algal meadows [23], because this multitude of gastropods requires a large number of ecological niches, which the bare seafloor hardly can provide to this extent. Even without fleshy algae being preserved, it is thus possible to conclude from the composition of the fauna that the soft bottom biotope was characterized by macroalgae. Many of the small invertebrates surely did not live on the sea bottom but on the macroalgae. This applies to the majority of the small gastropods, and also to many different small sessile suspensions feeders which used the algae as firm substrate, for instance byssate bivalves. Besides, the algal meadows also provided excellent living conditions for many different organisms which lived directly on the sea bottom or endobenthically. Organisms requiring firm or hard substrates, for fixation used stems of macroalgae, shells and other skeletal elements which were present on the seafloor in abundance, or shells of living invertebrates. The seafloor probably was covered by a heterogeneous, patchy mosaic of areas with dense vegetation of macroalgae and areas with thinner vegetation. Between the separate patches of macroalgae with their diverse fauna of small invertebrates, brachiopods, bigger suspension-feeding bivalves, crinoids and other animals preferably lived in the areas poorer in vegetation. In the algal meadows, species diversity was significantly higher than in the areas poorer in vegetation.

The brachiopod-dominated associations of relatively low diversity, which trophically are strongly dominated by suspension-feeders, on the other hand very probably represent relics of communities which lived on a seafloor poor in vegetation (Fig. 7). For organisms which required firm or hard substrates for fixation, appropriate substrates were provided by skeletal elements on the seafloor and by shells of other invertebrates.

Comparison
Wendt and Fürsich [16] presented a facies analysis of the Cassian Formation (Upper Ladinian - Lower Carnian) of the Southern Alps. The Anisian of Qingyan shares some of its aspects with regard to both facies and highly diverse and well-preserved fauna of small invertebrates. Among the facies described from the Cassian Formation, the closest resemblance exists with the shallow basinal facies of marly and clayey sedimentation with sporadical intercalations of marly limestone, which yields autochthonous soft bottom-algal meadow associations. However, the composition of the associations from Bangtoupo does not resemble that of the algal meadow associations rich in gastropods from the Cassian Formation [5]. Rather, it shows some similarities to the patch reef-dwelling associations from there, which are rich in gastropods as well. The possibility of the associations of Bangtoupo being relics of reef-dwelling faunas can be excluded, however, because no bioherms or biostromes occur during the whole Upper Anisian of Qingyan and potential reef-building organisms like corals, sponges or calcareous algae form only small parts of the associations. In spite of some similarities to the Cassian Formation, the conditions during the Upper Anisian of Qingyan thus were clearly different from those in the Southern Alps.

SYNOPSIS OF THE RECONSTRUCTED PALAEOSYNECOLOGICAL DEVELOPMENT (Fig. 8)

During the lower part of the Upper Anisian, the study area developed from a facies of deeper water with few pelagic organisms to shallow marine conditions in a protected habitat in the upper part of the slope (Bangtoupo section).

Initially, several different brachiopod-dominated communities of relatively low diversity became established on a seafloor probably poor in vegetation. Presumably due to further shallowing of the water and increasing degree of protection of the shallow marine habitat, colonization by macroalgae came about. Gradually, algal meadows of increasing density may have developed, until the seafloor constituted a heterogeneous, patchy mosaic of areas with dense vegetation of macroalgae and areas of thinner

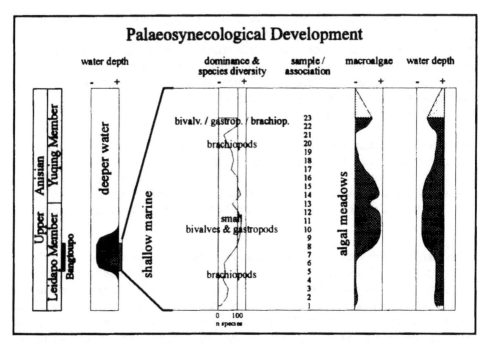

Figure 8. Synopsis of the reconstructed palaeosynecological development during the Upper Anisian of Qingyan.

vegetation. Parallel to this development, extremely diverse faunas of small invertebrates, especially bivalves and gastropods, began to flourish. For them the algal meadows provided ideal living conditions. The importance of the brachiopods had decreased very strongly in these associations (middle part of the Bangtoupo section). Later, another gradual and variable transition led back to strongly brachiopod-dominated communities of clearly lower diversity, which again lived on a seafloor poor in vegetation probably due to deepening of the water (upper part of the Bangtoupo section). Then another development of communities of higher diversity probably partly bound to macroalgal vegetation was initiated.

Finally, gradually increasing water depth terminated the flourishing of shallow marine benthic faunas. A transition took place from the shallow marine facies to deeper-water conditions with only few, mainly pelagic organisms, which lasted throughout the middle and upper part of the Upper Anisian. Only with the beginning of the Ladinian, water depth once more decreased and shallow marine faunas occurred.

Acknowledgements

Many thanks are due to Prof. J.H. Chen (Nanjing Institute of Geology and Palaeontology, Academia Sinica) for joint field work and various help during my stay in China and to Prof. Dr. F. Strauch (Geologisch- Paläontologisches Institut, Universität

Münster) for support in Germany. I am grateful to Dr. M. Bertling (GPI Münster) for critically reading the manuscript. Thanks also go to the Studienstiftung des deutschen Volkes for various financial support of this research project. This paper is part of a future doctoral thesis at the University of Münster.

REFERENCES

1. M.D. Brasier. An outline history of seagrass communities, *Palaeontology* 18:4, 681-702 (1975).
2. Bureau of Geology and Mineral Resources of Guizhou Province. Regional Geology of Guizhou Province, *Geol. Mem. Minist. Geol. Min. Res.* Ser. 1:7, 1-698. Geological Publishing House, Beijing (1982). [in Chinese with English summary]
3. Z.Q. Deng and L. Kong. Middle Triassic Corals and Sponges from Southern Guizhou and Eastern Yunnan, *Acta palaeont. Sinica* 23:4, 489-504 (1984). [in Chinese with English summary]
4. J.S. Fan. The main features of marine Triassic sedimentary facies in southern China, *Riv. Ital. Paleont. Strat.* 85:3-4, 1125-1146 (1980).
5. F.T. Fürsich and J. Wendt. Biostratinomy and palaeoecology of the Cassian Formation (Triassic) of the Southern Alps, *Palaeogeogr., Palaeoclimatol., Palaeoecol.* 22, 257-323 (1977).
6. Guizhou Stratigraphic and Palaeontologic Team (Ed.). [*Palaeontological Atlas of Southwest China, Guizhou Volume, Part 2: Carboniferous to Quaternary*]. Geological Publishing House, Beijing (1978). [in Chinese]
7. Z.A. He, H. Yang, and J.C. Zhou. The middle Triassic reef in Guizhou province, *Scientia Geol. Sinica* 1980:3, 256-264 (1980). [in Chinese with English summary]
8. E. Koken. Über triassische Versteinerungen aus China, *N. Jb. Mineral. Geol. Palaeont.* 1900:1, 186-215 (1900).
9. Nanjing Institute of Geology and Palaeontology, Academia Sinica (Ed.). [*Handbook of Stratigraphy and Palaeontology of Southwest China*]. Science Publishing House, Beijing (1974). [in Chinese]
10. Nanjing Institute of Geology and Palaeontology, Academia Sinica (Ed.). [*Fossil Bivalves of China*]. Science Publishing House, Beijing (1976). [in Chinese]
11. T. Sang (Ed.). *Atlas of lithofacies and palaeogeography of Guizhou (Mesoproterozoic to Triassic)*. Science and Technology Publishing House of Guizhou, Guiyang (1992). [in Chinese with English summary]
12. F. Stiller. Paläosynökologie einer oberanisischen flachmarinen Fossilvergesellschaftung von Leidapo, Guizhou, SW-China, *Münster. Forsch. Geol. Paläont.* 77, 329-356 (1995).
13. S. Sun, J.L. Li, H.H. Chen, H.P. Peng, K.J. Hsü, and J.W. Shelton. Mesozoic and Cenozoic Sedimentary History of South China, *Bull. Amer. Assoc. Petrol. Geologists* 73:10, 1247-1269 (1989).
14. H.Z. Wang, X.C. Chu, B.P. Liu, H.F. Hou, and L.F. Ma (Eds.). *Atlas of the palaeogeography of China*. Cartographic Publishing House, Beijing (1985). [in Chinese with English summary]
15. N.W. Wang. Palaeo-ecologic climatology of China with special reference to its plate tectonic implications for pre-Jurassic time, *J. Southeast Asian Earth Sci.* 3:1-4, 171-177 (1989).
16. J. Wendt and F.T. Fürsich. Facies Analysis and Palaeogeography of the Cassian Formation, Triassic, Southern Alps, *Riv. Ital. Paleont. Strat.* 85:3-4, 1003-1028 (1980).
17. D.Y. Xu (T.-Y. Hsu) and K. Chen. Revision of the Chingyen Triassic Fauna from Kueichou, *Bull. Geol. Soc. China* 23:3-4, 129-138 (1943).

18. Z.Y. Yang (T.-Y. Yang) and G.R. Xu (G.-Y. Xu). *Triassic Brachiopods of Central Gueizhou (Kueichow) Province, China.* Industrial Publishing House, Beijing (1966). [in Chinese with English summary]

19. H.F. Yin. Biostratigraphic Problems on the Triassic of Kueichow Province, China. A systematic Analysis of the Lithofacies and Faunal Ecology of Triassic of Kueichow, *Acta geol. Sinica* **42:3**, 289-306 (1962). [in Chinese with English summary]

20. H.F. Yin. On lithofacies and palaeoecology of the Triassic of Kweichow Province, China, *Scientia Sinica* **12:8**, 1169-1196 (1963).

21. H.F. Yin and E.L. Yochelson. Middle Triassic Gastropoda from Qingyan, Guizhou Province, China: 1-Pleurotomariacea and Murchisoniacea, *J. Paleont.* **57:1**, 162-187 (1983).

22. H.F. Yin and E.L. Yochelson. Middle Triassic Gastropoda from Qingyan, Guizhou Province, China: 2-Trochacea and Neritacea, *J. Paleont.* **57:3**, 515-538 (1983).

23. H.F. Yin and E.L. Yochelson. Middle Triassic Gastropoda from Qingyan, Guizhou Province, China: 3-Euomphalacea and Loxonematacea, *J. Paleont.* **57:5**, 1098-1127 (1983).

Proc. 30th Int'l. Geol.. Congr., Vol. 12, pp. 161-171
Jin and Dineley (Eds)
© VSP 1997

Shell Preservation of Pleistocene Freshwater Bivalvian Mollusc from China

IWAO KOBAYASHI[1], QINGSHAN YU[2]

[1] *Department of Geology, Faculty of Science, Niigata University, Nino-cho, Ikarashi, Niigata 950-21, Japan*
[2] *Institute of Geology, Chinese Academy of Geological Sciences, Baiwanzhuang Road, China*

Abstract

Internal shell microstructures of molluscan fossils are important for paleontological and geological studies. The specimen, *Lamprotula*, of Early Pleistocene age, has well preserved internal microstructure which could be examined by the use of a scanning electron microscope. The shell structure is composed of a periostracum and calcareous shell layer. The latter is divided into three calcareous shell layers, namely outer, middle and inner. The outer consists of an aragonite layer which is made of prisms surrounded by preserved organic frameworks. The middle and inner layers consist of a nacreous film which is subdivided into pillow (lenticular) and sheet nacreous layers. The biomineral composition, aragonite is determined by the use of a X-ray diffractometer. A recent specimen was also studied for comparison. One important observation is that the inner calcareous layer is made of alternative layers of pillow and sheet nacreous structures.

Keywords: Lamprotula, Pleistocene, shell structure, nacreous layer, aragonite prismatic layer, aragonite, fossilization

INTRODUCTION

A nacreous layer of molluscs is an industrially and scientifically important shell structure, because pearl is made of this substance. It is widely developed in the calcareous shell layer of gastropods, pelecypods and cephalopods, Phylum mollusca [1, 3, 12, 6]. Even in molluscan fossils, the remains of this layer are well preserved. In this report, fossil and recent specimens collected from China were selected for examination by the use of optical and scanning electron microscopes. The purpose is to examine the preservation of internal shell structures, and to discuss the paleontological and paleoenvironmental significance.

MATERIAL AND METHOD

Material
Fossil material examined (Fig. 1A) is a Pleistocene bivalve from China. The specimen

Figure 1. Materials examined. A: *Lamprotula* aff. *paihoensis*, B: *Lamprotula* sp. Scale: 1cm.

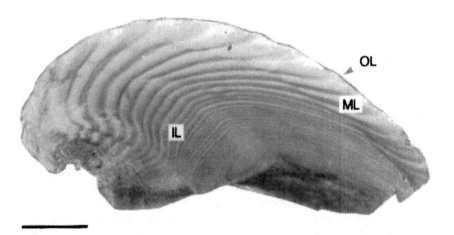

Figure 2. Radial polished plane of *Lamprotula* aff. *paihoensis*. OL: aragonite prismatic structure, ML: pillar nacreous structure partly with sheet nacreous structure, IL: pillar and sheet nacreous structures. Scale: 1cm.

Figure 3. Optical microscopic photographs of radial thin sections. A-D:*Lamprotula* aff. paihoensis, A: outer layer, aragonite prismatic st., B: middle layer, pillar nacreous st., C: inner layer, pillar and sheet nacreous st., D: inner layer, sheet nacreous st., E-H: *Lamprotula* sp., E-F: outer layer, aragonite prismatic st., G: middle layer, pillar nacreous st., H: inner layer, sheet nacreous st. Scale: 0.1mm.

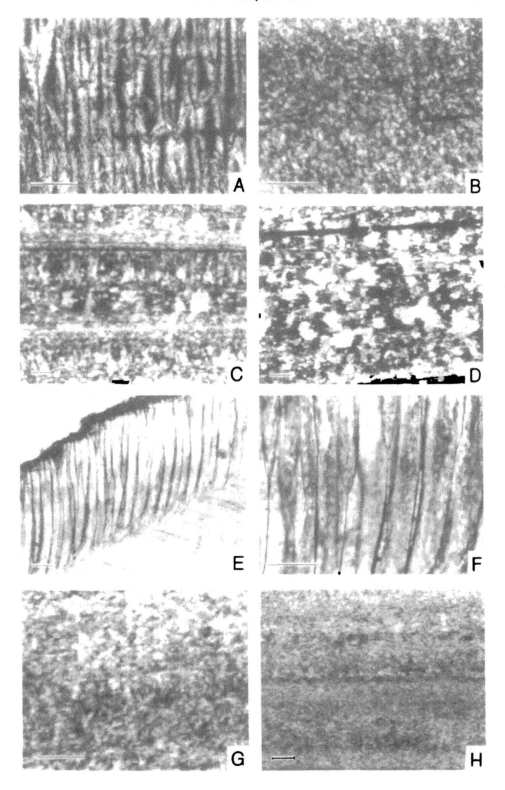

is *Lamprotula (Parunio)* aff. *Paihoensis*, obtained from the Early Pleistocene Shanmen Formation in Shanmenxia, Hena Province, China. And the other recent specimen (Fig. 1B), *Lamprotula* sp. was collected from Lake Taihu, China.

Method

Morphological observations were carried out under optical and scanning electron microscopes. Thin sections for optical microscopy were prepared by a method similar to making a thin section of rocks. Fractured and polished/etched surfaces of a shell were prepared for scanning electron microscopy. A polished surface was etched by 0.5mol EDTA solution (pH7.4) for 10 to 20 minutes. Shell mineral composition was examined by the use of a X-ray diffractometer.

RESULT

General Feature

The fossil specimen examined shows very well preservation in outview. A shell is partly covered by the remains of an organic periostracum which is light brown in color. A calcareous shell is white in color, partly brownish probably due to iron materials, moreover the inner shell surface has a weak luster of pearl. Original internal shell structures also are very well preserved.

Layer Structure

The shell of ancient *Lamprotula* is composed of thick calcareous shell layers covered by a light brownish thin film of periostracum on the outer shell surface. A shell in general consists of periostracum and calcareous shell layers. The lattter is composed of outer, middle and inner, and pellucid components [5, 6, 12].

Fig. 2A shows the radial polished plane of this specimen and Fig.2B explains the photograph of Fig. 2A. A periostracum is not well observed by this method. A calcareous shell layer is very thick, about 30 mm in thickness near the apex area. Three layers except for a pellucid layer can be discriminated from each other. The boundary of middle and inner layers is not well defined near the apex area. The outer layer is very thin less than 1mm in thickness and only partly preserved. This is white in color and opaque. The middle and inner layers are whitish opaque and translucent. The growth texture is very distinctive in alternative color patterns, such as whitish opaque and translucent zones. The thickness of middle layer is about 5mm near the ventral area. The inner layer is 20 mm in the central part. The pellucid layer cannot be observed under microscopy.

The recent *Lamprotula* specimen for comparison has a blackish brown periostracum which wholly covers an outer shell surface except for the apex area. On the other hand, the periostracum of the fossil specimen is light brownish in color. The shell of the recent specimen is also composed of a periostracum and calcareous shell layers which are divided into three layers and a pellucid layer. The periostracum observed under optical microscopy is black in color and 0.2mm in thickness. The outer calcareous layer

Figure 4. Scanning microscopic photograph of aragonite prismatic structure. A: *Lamprotula* aff. *paihoensis*, B: *Lamprotula* sp., P: prism, O: organic sheath. Scale: 10 microns.

is whitish opaque. The middle and inner calcareous layers are translucent with the lustre of pearl. The total thickness of the shell is less than that of the fossil specimen. The inner layer of the recent specimen is not thicker compared with the fossil one.

Morphological Shell Type

The shell microstructure of the fossil specimen has retained their original textures very well. The outer calcareous shell layer is composed of aragonite prismatic structure. The middle calcareous shell layer is composed mainly of pillow nacreous structure. The inner calcareous shell layer is composed of pillow and sheet nacreous structures. Similarly, the shell structure of the recent *Lamprotula* consists of the same morphological types except for the inner calcareous shell layer. The shell layer of the recent specimen is made only of sheet nacreous structure.

Aragonite Prismatic Structure

This structure [12, 6, 11] is made of subpolygonal prisms electing perpendicular to the shell surface. Each prism is composed of abundant elongate crystallites which are arranged in conical shape. The fossil *Lamprotula* has the same structure as the recent specimen. Prisms and crystallites are very well preserved in original form. The diameter of each prism is about 30 microns in average. Figs. 3A, 3E, 3F, 4 show the optical and scanning microscope photographs of thin sections and polished/etched planes of fossil and recent specimens.

Pillow (Lenticular) Nacreous Structure

This structure [12, 13, 11] is composed of vertically piled tablets forming prisms and arranged side by side (Fig. 5). Each tablet is polygonal in shape. The size of each tablet of the fossil *Lamprotula* is about 15 microns in diameter and about 3 microns in thickness. Figs. 3B, 3C, 3G, 6, show the photographs of thin sections and polished/etched planes of fossil and recent specimens. Some tablets have several sectors [10] dividing radially from the central part.

Sheet Nacreous Structure

This structure [12, 6] is composed of the accumulation of horizontally arranged tablets (Fig. 4B). Each tablet is polygonal in form. The size of tablets of the fossil *Lamprotula* is about 20 - 30 microns in diameter. Figs. 3C, 3D, 3H, 7 show the photographs of thin sections and polished/etched planes of fossil and recent specimens.

Growth Structure

The growth structure of the outer layer of fossil *Lamprotula* is not well-revealed under the microscope. On the other hand, the middle and inner layers are clearly observed, but the growth markings are not different from ordinary growth patterns related to day and tide. Under the observation of the polished plane (Fig. 2), the growth structure is seen as alternative color patterns of whitish opaque and translucent zones in the middle and inner layers. Especially in the inner layer of the fossil specimen, both zones are very clear and alternate. The opaque zone is a little wider than the translucent zone in the radial section. The thickness of one set of both zones is 2 - 0,5 mm and is thinner

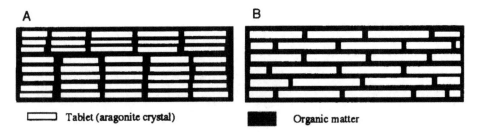

Figure 5. Two types of nacreous structure. A: pillar (lenticular) nacreous structure. B: sheet nacreous structure.

Figure 6. Scanning microscopic photograph of radial polished/etched plane of pillar nacreous structure. *Lamprotula* aff. *paihoensis.* Scale: 10 microns.

Figure 7. Scanning microscopic photograph of radial polished/etched plane of sheet nacreous structure *Lamprotula* aff. *paihoensis.* Scale: 50 microns.

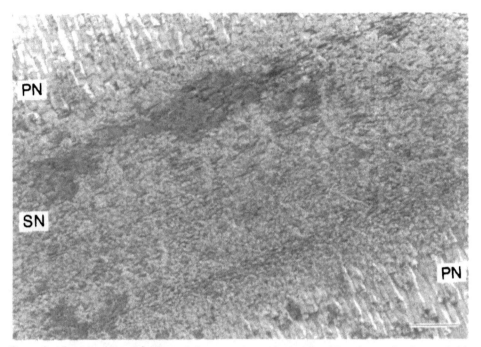

Figure 8. Alternative layers of pillar and sheet nacreous structures. Scanning microscopic photograph of a radial section of *Lamprotula* aff. *paihoensis*. PN: pillar nacreous structure, SN: sheet nacreous structure Scale: 100 microns.

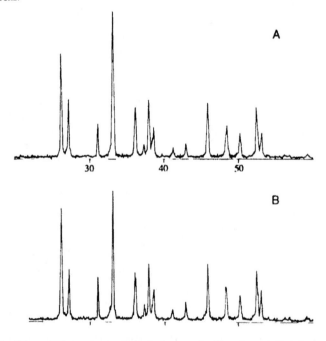

Figure 9. Result of X-ray diffractmeter analysis. A: *Lamprotula* aff. *paihoensis*, B: *Lamprotula* sp.

towards the inner surface. Next, it is important that each zone consists of different morphological types of shell structure. The whitish opaque zone is composed of pillow nacreous structure, and on the other hand, the translucent zone is of sheet nacreous layer. Fig. 8 shows shell structures of both zones under scanning electron microscopy. One set of both zones was counted about 40 in number. The alternative layers are very characteristic and peculiar. This texture is not seen in the recent specimen.

Biomineral
Biominerals of fossil and recent *Lamprotula* specimens were examined by the use of a X-ray diffractometer. The result (Fig. 9) indicates that all biominerals are aragonite. From the data, it is thought that elongate crystallites of aragonite prismatic structure and tablets of nacreous structure are made of aragonite.

Organic matrix
Aragonite prismatic structure has strong organic sheaths around prisms. In this examination, the outer calcareous layer of the fossil *Lamprotula* consists of aragonite prismatic structure with the remains of organic sheaths around prisms. Fig.4A of scanning electron microscopy shows some membraneous materials around prisms. These membranes are remains of organic sheaths, but, probably altered from the original organic materials. Of course, organic sheaths of the aragonite prismatic layer of the recent *Lamprotula* are clearly observed (Fig. 4B).

DISCUSSION

Preservation of Shell Microstructure
In most cases, the original aragonitic composition of shell fossils collected from Mesozoic and Paleozoic strata was very often replaced by calcite in beds. However, this Pleistocene specimen has retained its original biomineral component as well as shell microstructure. Moreover, altered organic matters are seen in the outer shell layer. Japanese Late Miocene specimens also have an original structure of internal shell structures [7]. Shell minerals and structures of Japanese Early Miocene specimens were mostly altered. Cretaceous ammonites in nodules from Hokkaido, Japan, have often been preserved with their original structure of a nacreous layer.

Shell Structure
As the shell of the fossil *Lamprotula* is well preserved, we can very easily determine morphological type by the use of optical and scanning microscopes. The result is basically same to the shell structure of recent Unionidae [12, 6]. Aragonite prismatic, pillow nacreous and sheet macreous structures are well preserved. The unit component such as elongate crystallites and tablets remains in original form without any change. However, the diagenetical change of chemical component has not been examined. Inner periostracums in a nacreous layer are formed in many species of Unionidae, but they are not observed from this fossil *Lamprotula*. This maybe is related to the development of peculiar periodical banding structures.

Biomineral

Aragonite prismatic and nacreous structures are made of aragonite crystals. The fossil and recent specimens are the same mineral, aragonite. This indicates original materials and that the thermal condition in a bed was not high.

Organic Matrix

Organic remains of a fossil from Cenozoic and Mesozoic strata were found [3, 4]. They were also observed in the aragonite prismatic layer of the fossil *Lamprotula*. They are organic sheath remains of the aragonite prismatic structure. Generally, organic matters decompose earlier than calcareous parts except for under acidic environment. The remain of this organic sheaths was probably altered into stable materials. In many cases [8], the organic membranes are altered by iron materials. The chemical analysis of sheaths needs for the determination of component.

Periodical Banding

In this fossil specimen, especially inner layers have a periodical banding structure which is composed of two different nacreous structures. Both structures are similar in structural aspect, but the growth pattern is a little different in each. It is thought from observations that the pillow nacreous structure grows faster in the vertical direction, and the sheet nacreous structure towards horizontal direction. If this is correct, these periodical bandings indicate difference in growth speed. The formation of bandings is related to difference in growth rate and indicates the existence of a repeated change such as annual or seasonal one [1, 9]. The meaning of these structures is significant for environmental problems.

Phylogenetic Problem

In Unionidae, in general the outer calcareous layer is composed of an aragonitic prismatic layer similar to a normal prismatic layer. The fossil *Lamprotula* also has the same structure as the recent *Lamprotula* except for one phenomenon. The middle and inner calcareous layers are composed of a nacreous structure, moreover these nacreous layers are of two types. In general, the middle calcareous layer consists of pillow nacreous structure and the inner layer does of sheet nacreous structure. This fossil *Lamprotula* is, thus, different from the shell structure of general Unionidae in one point.

Acknowledgement

We would like to express our thanks to Prof. J. Akai, Niigata University, for X-ray analyses.

REFERENCES

1. R.M. Barker. Microtextural variation in pelecypod shells, *Malacologia* **2**, 69-86 (1964).

2. O.B. Boggild. The shell structure of the mollusks, *D. Kgl. Danske Vidensk. Selsk. Skr., Naturv., Ser. 9,* **2,** 232-325 (1930).

3. C. Gregore. Topography of the organic components in mother-of-pearl, *J. Biophys. Biochem Cytol.* **3,** 797-808 (1957).

4. G.R. Clark II.,1993, Physical evidence for organic matrix degradation in fossil Mytilid (Mollusca: Bivalvia) shell, In I. Kobayashi, H. Mutvei and A. Sahni (Eds.): *Structure, Formation and Evolution of Fossil Hard Tissues.* Tokai Univ. Press, Tokyo, 73-79 (1993).

5. I. Kobayashi. Intriduction to the shell structure of bivalvian molluscs, *Earth Science* **73,** 1-12 (1964).

6. I. Kobayashi. Internal shell microstructure of recent bivalvian molluscs, *Sci. Rep. Niigata Univ., Ser.E, Geol. Mineral.* **2,** 27-50 (1971).

7. I. Kobayashi. The change of internal shell structure of Anadara ninohensis (Otuka) during the shell growth, *Jour. Geol. Soc. Japan* **82,** 441-447 (1976). (in Japanese with English abstract)

8. I. Kobayashi. Change of inorganic matters in the course of fossilization -Inorganic component and depositional style of iron sulphide mineral in fossil teeth-, *Contrib. Dept. Geol. Mineral. Niigata Univ.* **7,** 1-25 (1992). (in Japanese with English abstract)

9. R.A. Lutz and D.C. Rhoads, D. C. Growth patterns within the molluscan shell an overview, In D.C. Rhoads and R.A. Lutz (Eds.): *Skeletal Growth of Aquatic Organisms.* Plenum Press, New York, 203-254 (1980).

10. H. Mutvei, H. The nacreous layer in molluscan shells, In M. Omori and N. Watabe (Eds.): *The mechanisms of Biomineralization in Animals and Plants.* Tokai Univ. Press, Tokyo, 49-56 (1980).

11. H. Nakahara. An electron microscope study of the growing surface of nacre in two gastropod species, *Turbo cornutus and tegula pfeifferi. Venus (Jap.J.Malac.)* **38,** 205-211 (1979).

12. J.D. Taylor, W.J. Kennedy and A. Hall. The shell structure and mineralogy of the bivalvia. Introduction, Nuculacea-Trigonacea, *Bull. Brit. Mus., Nat. Hist., Zool., Supple.* **3,** 1-125 (1969).

13. S.W. Wise, S.W. Microarchitecture and mode of formation of nacre (mother-of-pearl) in Pelecypods, Gastropods, and Cephalopods. *Eclogae Geol. Helv.* **63,** 775-797 (1970).

Proc. 30* Int'l. Geol. Congr., Vol. 12, pp. 172-181
Jin and Dineley (Eds)
© VSP 1997

First discovery of Tertiary Fishes (Teleostei) from Tarim Basin, Xinjiang, China

ZHANG JIANGYONG

Institute of Vertebrate Paleontology and Paleoanthropology, Chinese Academy of Sciences, P. O. Box 643, Beijing, China

Abstract

New specimens from the Tarim Basin, Xinjiang, China were identified as Clupeidae gen. indet. because they have the following synapomorphies of clupeomorph fishes (Grande 1985): 1. Presence of one or more abdominal scutes, each of a single element which crosses the ventral midline of the fish (synapomorphy of Clupeomorpha); 2. An autogenous first hypural (synapomorphy of division 2 of Clupeomorpha); 3. Loss of the beryciform foramen (synapomorphy of Clupeiformes); 4. Fusion of the first uroneural with the first preural centrum (synapomorphy of Clupeoidei); 5. Reduction in relative size of the first ural centrum(synapomorphy of Clupeoidei); 6. Presence of two long, rodlike postcleithra (synapomorphy of Clupeidae). It is difficult to make a detailed comparison with other genera of Clupeidae due to the poor preservation of the fish from the Tarim Basin.

Keywords: Teleostei, Tarim Basin, China, Tertiary

INTRODUCTION

The Tarim Basin is the largest interior basin in China, located in southwestern Xinjiang, northwestern China and encircled by the Tianshan Mountains on the north, the Kunlun and Altun mountains on the south. The basin is nearly rhombic in plan, with an area of about 560,000 km².

In 1992, I worked in the Tarim Basin with the expedition team from Nanjing Institute of Geology and Paleontology under the auspices of the program "The Stratigraphy and Paleontology of the Tarim Basin". This program belongs to the project "The Oil-Gas Resources of Tarim Basin" which is part of the 8th Five-year Plan of Science and Technology of China. The fishes were collected in Xiyanshuigou, about 80 km to the southwest of the county city of Baicheng (Fig. 1).

The strata of the Suweiyi Formation in which the fishes were collected consist of red-brown mudstones, sandy mudstones, the alternation of gray, yellow-brown sandstones and siltstones, and layers of gypsum and halite. In addition to fishes, abundant ostracodes (*Cyprideis, Mediocypris* and *Ilyocypris*) and charophytes (*Dongmingochara* and *Chara*) were also found in this locality (Lu Hui-nan and Luo Qi-xin 1990, Fang Zong-jie *et al.* in press).

No consensus of opinion has been reached on the geological age of Suweiyi Formation so far. It was considered to be Late Eocene to Oligocene by Lu Hui-nan and Luo Qi-xin (1990) and Bureau of Geology and Mineral Resources of Xinjiang Uygur Autonomous Region (1993), Early Miocene by Fang Zong-jie *et al.* (in press), Kang Yu-zhu (1987) and Zhou Zhi-yi and Chen Pi-ji (1990).

The specimens are poorly preserved and most of them are detached bones. The fishes belong to the Clupeiformes and the Perciformes respectively, but only the Clupeomorph fish are studied in the present paper. Since these are the first teleost fish remains from the Tarim Basin, it is worthwhile to report this discovery.

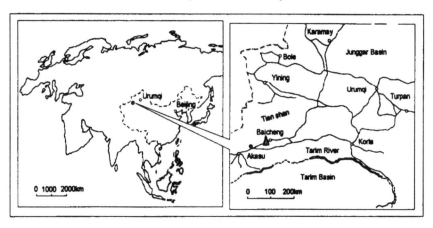

Figure 1. Map of the locality of Tertiary fishes in Baicheng, Xinjiang, China. (▲ Xiyanshigou)

MATERIAL AND METHODS

The specimens studied were collected by me in 1992 and deposited in IVPP (Institute of Vertebrate Paleontology and Paleoanthropology, Chinese Academy of Sciences). All specimens were mechanically prepared. The exclusively fossil taxa are marked with daggers preceding their names. The drawings were executed under a Wild 308700 microscope with camera lucida attachment.

ANATOMICAL ABBREVIATIONS

H1-2, hypural 1-2; hsPU3, haemal spine on PU3; nsPU3, neural spine on PU3; p.an, anterior process of the neural or haemal arch; pfmx, palatine condyle on Maxilla; Ph, parhypural; PU, preural centrum; PU1+U1+un1+Ph, the first preural centrum + the first ural centrum + the first uroneural + parhypural; un1-2, uroneural 1-2.

SYSTEMATIC PALEONTOLOGY

Superorder Clupeomorpha Greenwood *et al.* 1966
Order Clupeiformes (*sensu* Grande 1985)
Suborder Clupeoidei (*sensu* Grande 1985)
Family Clupeidae (*sensu* Grande 1985)
Genus indeterminatum

Material: a caudal skeleton, part of a deformed head and 54 scattered bones and scales, IVPP 10774. 1-56.
Locality: Xiyanshuigou, Baicheng, Xinjiang, China.
Occurrence: Suweiyi Formation.
Age: Late Eocene to Early Miocene.

Description
Little is known of the braincase except the supraoccipital (Fig. 2,A) which is nearly round with a small occipital crest.

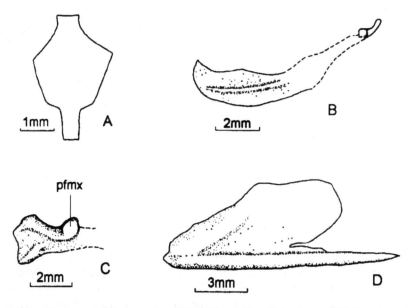

Figure 2. Clupeidae gen. indet. **A**, supraoccipital; **B**, maxilla; **C**, head of maxilla; **D**, dentary.

Jaws: The maxilla (Fig. 2, B-C; Fig. 3,A-B) is an elongate bone with a large and flattened head. There is a strong, rounded condyle on the dorsal surface of the maxilla which probably articulated with the head of palatine. No specially developed cranial condyle is found. A prominent ridge is present along the middle part of the maxilla and a single row of tiny teeth along the oral margin (Fig. 2,B; Fig. 3,B).There are two supramaxillae (V10774.1). The posterior supramaxilla (Fig. 4,A) is large with a long process extending above the anterior supramaxilla. The anterior supramaxilla (Fig. 3,C) is long and slender.

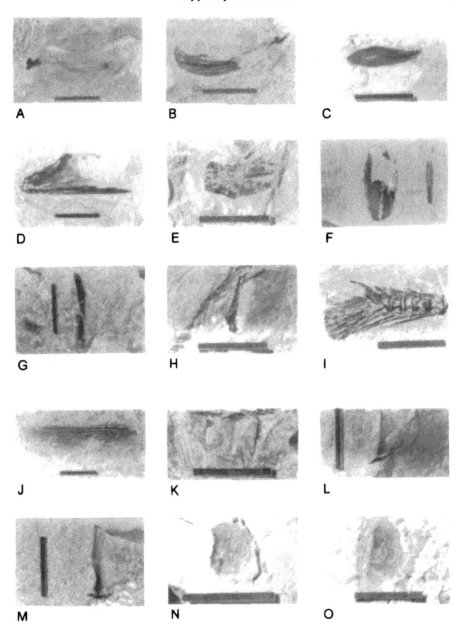

Figure 3. Clupeidae gen. indet. **A-B**, maxilla (**A**, V10774.5; **B**, V10774.6); **C**, anterior supramaxilla (V10774.30); **D**, dentary (V10774.8); **E**, anterior ceratohyal (V10774.29); **F**, orpercle (V10774.15); **G**, postcleithra (V10774.18); **H**, neural arch and neural spine (V10774.38); **I**, caudal skeleton (V10774.45); **J**, caudal fin (V10774.46); **K-M**, abdominal scutes (**K**, V10774.25b; **L**, V100774.39; **M**, V100774.42); **N-O**, scales (**N**, V100774.48; **O**, V100774.51). Scale bars equal 4mm.

A prominent coronoid process is present in the lower jaw. The dentary (Fig. 2,D; Fig. 3, D) has a moderately deep symphysis with a sharply rising oral margin and nearly straight ventral margin.

No teeth are found on dentary. The articular surface for the quadrate of the angulo-articular is deeply excavated (Fig. 4,B, V10774.11). The coronoid process of the angulo-articular rises gently. There is a depression on the lateral surface of the bone. The mandibular sensory canal runs through the postero-ventral part of the angulo-articular.

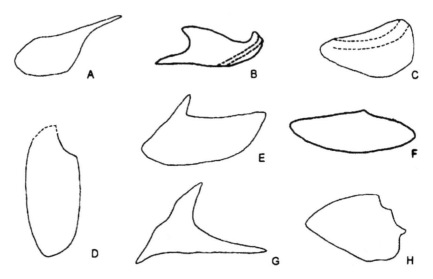

Figure 4. Clupeidae gen. indet. **A**, posterior supramaxilla; **B**, angulo-articular; **C**, the third infraorbital; **D**, orpercle; **E**, subopercle; **F**, interopercle; **G**, post-temporal; **H**, coracoid.

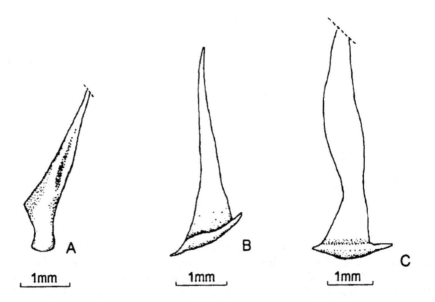

Figure 5. Clupeidae gen. indet. **A**, neural arch and neural; **B-C**, abdominal scutes.

Because of poor preservation, nothing can be said about the dorsal, pelvic and anal fins, except a pterygiophore in specimen V10774.18 which is pike-like in shape.

Caudal skeleton and fin: A nearly complete caudal skeleton (Fig. 3,I; Fig. 6) can be seen in specimen V10774.45. The preural vertebrae gradually reduce in size posteriorly. The last three haemal spines are long and flattened. The neural and haemal arch fused with their respective preural centra and have anterior processes. Posterior to the second preural centrum is a compound element fused by the first preural centrum, first ural centrum, uroneural 1 and parhypural (PU1+U1+un1+Ph). The second ural centrum is not found. A long, small bone with a pointed anterior end posterior to the compound element is uroneural 2. The posterior part of the uroneural 2 is missing.

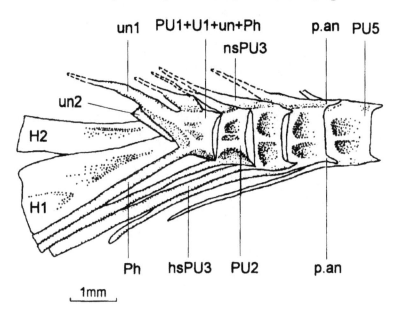

Figure 6. The caudal skeleton of Clupeidae gen. indet.

Two hypurals are preserved. Hypural 1 is larger than hypural 2 and is separated from the compound element. It is hard to say whether the hypural 2 is fused with or separated from the compound element because its proximal end is covered by the uroneural 2.

A partial caudal fin (Fig. 3,J) can be seen in specimen V10774.46 including one unbranched, six branched principal and one procurrent rays. The inner rays, minus the innermost ones, are much shorter than the out ones indicating that the caudal fin is deeply forked.

Squamation: Six abdominal scutes (Fig. 3,K-M; Fig. 5,B-C) are preserved in specimen V10774.25 in ventral view, one in each of the specimens V10774.39 and V10774.42 in lateral view. The scutes bear a prominent keel and a long ascending arm, similar to those found in many clupeiforms. No dorsal scute was found.

The scales (Fig. 3,N-O; Fig. 7) are thin and cycloid, with dense circulii and two or three radii vertically arranged. The scales are usually broken along the radii.

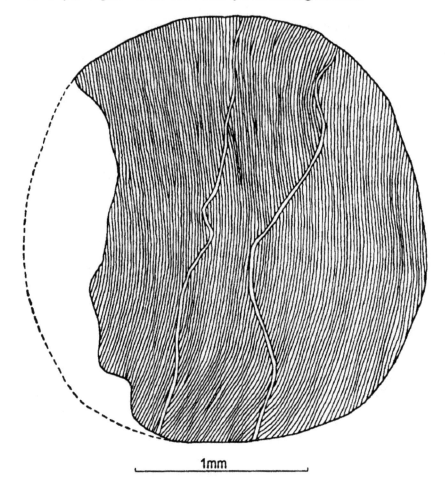

1mm

Figure 7. Clupeidae gen. indet. A scale showing the nearly vertically arranged circuli.

RELATIONSHIPS

From the defining characters of the clupeomorph fishes given by Grande (1985), characters 12, 13, 17, f and m are absent, characters 2-3, 5-10, 18-22 are unclear, and the following uniquely derived characters are present in the specimens from Baicheng: 1. Presence of one or more abdominal scutes, each of a single element which crosses the ventral midline of the fish (synapomorphy of Clupeomorpha); 4. Hypural 2 fused with the first ural centrum (the fusion of hypural 2 with the first ural centrum is unclear in the fish from Baicheng), and an autogenous first hypural (synapomorphy of division 2 of Clupeomorpha); 11. Loss of the beryciform foramen (synapomorphy of Clupeiformes); 14. Fusion of the first uroneural with the first preural centrum (synapomorphy of Clupeoidei); 15. Reduction in relative size of the first ural centrum (the first ural centrum is fused with the first preural centrum in the fish from Baicheng, but the posterior part of the compound element corresponding to the first ural centrum is relatively small)(synapomorphy of Clupeoidei); 23. Presence of two long, rodlike postcleithra (synapomorphy of Clupeidae). These characters strongly suggest that the fish from Baicheng is a member of the Clupeidae.

Character 17 (separation of the parhypural from the first preural centrum) was considered to be a synapomorphy of the Clupeoidei (Grande 1985), but the parhypural is fused with the first preural centrum in the fish from Baicheng. This fusion is also present in other two clupeoids (*Dussumieria, Etrumeus*)(Grande 1985). Because most clupeoids have no such a fusion, it is considered to be derived for the fish from Baicheng, *Dussumieria*, and *Etrumeus*.

Within the Clupeidae, the interrelationships of the subgroups is unclear. The subfamilies Dussumieriinae and Pellonulinae are monophyletic (Grande 1985). Dussumieriinae which lack nonpelvic abdominal scutes are unique in having a peculiar, unkeeled, W-shaped pelvic scute and Pellonulinae missing anterior supramaxilla. The fish from Baicheng has abdominal scutes and two supramaxilla. Therefore, it is difficult to place the fish in either of these two subfamilies.

The monophyly of the other three clupeid subgroups (Alosinae, Dorosomatinae and Clupeinae) is doubtful. No valid osteological characters could be found for them (Grande 1985). The only known fish in which the first preural centrum fused with the first ural centrum as in the fish from Baicheng among Alosinae, Dorosomatinae and Clupeinae is *Clupeonella cultiventris* from the Black Sea (Grande 1985). The fossil *Clupeonella* was found from the Lower Miocene of Germany, and the Late Tertiary of Armenia and Russia (Danil'chenko 1980, Woodward 1901). Because no specimen of fossil *Clupeonella* was examined, whether it has the fused caudal centra or not is unknown to the author. Outside the three subgroups in Clupeomorpha, the fusion also occurs in some pellonulines, dussumieriines and engraulids (Grande 1985).

Because of lacking meristic and morphological information due to poor preservation, the fish from Baicheng can be only identified to family level (Clupeidae gen. indet.).

The fish from Baicheng is clearly distinct from the other known fossil clupeids from China. *Knightia bohaiensis* (Chang Mee-mann *et al.* 1985), from Eocene transitional deposits of Shandong Province, has isolated first preural and first ural centra, autogenous parhypural and neural and haemal arches, unenlarged first hypural, scales with a comb-like posterior margin. Only one poorly preserved specimen, †*Knightia yuyanga* (Liu Hsien-T'ing 1963,), from Late Cretaceous lacustrine deposits of Hubei Province (Lei yi-zhen 1987), is prominently different from the fish of Baicheng in that its abdominal scutes have long posterior spines and much longer ascending arms.

AGE AND ENVIRONMENT

The discovery of the fish in Baicheng is important because it is the first time that teleost fish were found in the Tarim Basin. The age of the Suweiyi Formation has long been disputed. It was considered to be Late Eocene to Early Miocene. According to the newly found ostracodes (*Cyprideis*), Fang Zongjie *et al.* (in press) argued that the age is probably Early Miocene. Both the fish from Baicheng and *Clupeonella cultiventris* have fused first preural and first ural centra. If further study shows that the fossil *Clupeonella* has such a fusion and other similarities to the fish from Baicheng, the age of the Suweiyi Formation may be correspond to that of the deposits with *Clupeonella* (i.e. Early Miocene or Late Tertiary).

The environment in which the Suweiyi Formation was deposited was considered to be lagoonal (Zhou Zhiyi and Chen Piji 1990) or a saline lake (Fang Zongjie *et al.* in press) according to the foraminifera and ostracodes associated with the fish. The fishes found in Baicheng belong to Clupeiformes and Perciformes respectively. Most clupeids and perciforms inhabit marine or brackish water. No typical freshwater fishes, such as cyprinids, were found in this locality. This fossil fish assemblage also suggests a lagoon or saline lake environment, probably in someway connected to the sea.

Acknowledgments

I am indebted to Professor Chang Meemann of IVPP, Beijing, for criticizing the manuscript. I thank the expedition team from Nanjing institute of Geology and Paleontology for their help during the field work. Thanks also go to Professor Zhou Jiajian from IVPP for her help in photographing the specimens.

REFERENCES

1. Bureau of Geology and Mineral Resources of Xinjiang Uygur Autonomous Region. Geological Memoirs Series 1:32. *Regional Geology of Xinjiang Uygur Autonomous Region*. Geological Publishing House, Beijing. 1-841 (1993) (in Chinese with English summary).
2. Chang Meemann, Zhou Jiajian and Qin Derong. *Tertiary Fish Fauna From Coastal Region of Bohai Sea*. Memoirs of Institute of Vertebrate Paleontology and Paleoanthropology, Academia Sinica. No. 17, 1-60 (1985) (in Chinese with English summary).
3. Danil'chenko, P. G. Otryad Clupeiformes (Order Clupeiformes). In *Iskopayemyye kostistyye ryby SSSR*. Novitskaya, Larisa Illarionova (eds.), Tr. paleontol. Inst. 178, 7-26 (1980).
4. Fang Zongjie *et al. The Stratigraphy and Paleontology of the Tarim Basin* (in press).
5. Grande, L. Recent and fossil clupeomorph fishes with materials for revision of the subgroups of clupeoids. *Bulletin of the American Museum of Natural History* 181:2, 231-372 (1985).
6. P.H. Greenwood, D.E. Rosen, S.G. Weitzman, and G.S. Myers. Phyletic studies of teleostean fishes, with a provisional classification of living forms. *Bulletin of the American Museum of Natural History* 131:4, 339-456 (1966).
7. Kang Yuzhu. A discussion on the relationship between the lithofacies in the palaeogeography during the Meso-Cenozoic era and the oil-gas in the Tarim Basin. In *the Collection Works on*

Meso-Cenozoic Geology, The Geological Society of Beijing, Geological Publishing House, Beijing. 365-377 (1987) (in Chinese with English summary).

8. Lei Yizhen. Biostratigraphy of the Yangtze Gorge Area (5): *Cretaceous and Tertiary*. Geological Publishing house, **104**: 191-194; Beijing (1987) (in Chinese with English summary).

9. Liu HsienT'ing. The discovery of double-armoured herrings from Itu, Hupei. *Vertebrata PalAsiatica* **7:1**, 31-37 (1963) (in Chinese with English summary).

10. Lu Hui-nan and Luo Qi-xin. *Fossil Charophytes from the Tarim Basin, Xinjiang*. Scientific and Technical Documents Publishing House, Beijing. 1-216 (1990) (in Chinese with English summary).

11. Woodward, A. S. *Catalogue of the fossil fishes in the British Museum*. Vol. **4**: Containing the actinopterygian Teleostomi of the suborders Isospondyli (in part), Ostariophysi, Apodes, Percesoces, Hemibranchii, Acanthopterygii and Anacanthini. London, Taylor and Francis, 1-636 (1901).

12. Zhou Zhi-yi and Chen Pi-ji. *The Biostratigraphy and Geological Evolution of the Tarim Basin*. Science Press, Beijing. 1-366 (1990).

Proc. 30ᵗʰ Int'l. Geol. Congr., Vol. 12, pp. 182-192

Jin and Dineley (Eds)

© VSP 1997

Ultra-microfossils in Manganese Stromatolites from the East Pacific Ocean

BIAN LIZENG[1], LIN CHENGYI[2], ZHANG FUSHENG[2], DU DEAN[2], CHEN JIANLIN[3], SHEN HUATI[3] and HAN XIQIU[3]

[1] *Department of Earth Sciences, Nanjing University, Nanjing 210093 P. R. China*
[2] *Center for Materials Analysis, Nanjing University, Nanjing 210093 P. R. China*
[3] *Second Institute of Oceanography, State Oceanic Administration, Hangzhou 310012 P.R. China*

Abstract

Pelagic manganese nodules from the East Pacific Ocean have been studied by electron microscopy, paleontology and mineralogy. The results show that they are composed of core and coatings, and the coatings are manganese stromatolites. Based on lamina shape, column shape and its branching and coalescing, two forms of stromatolites, namely *Minima* and *Admirabilis*, have been identified. They can be seen in the coatings of smooth and knobbly nodules respectively. It has been revealed from TEM studies that these manganese stromatolites have been constructed by the ultra-microfossils. According to the morphology of the hyphae two species of ultra-microfossils, *Miniactinomyces chinensis* and *Spirisosphaerospora pacifica*, have been identified. *Miniactinomyces chinensis* and *Spirisosphaerospora pacifica* were sessile benthic microorganisms living in completely darkness on rocky or other solid substrate. The scattered occurrence, presence of various cores, radial growth and some other features can be explained by biogenic formation of the pelagic manganese nodules. They are not a product of either crystallization or sedimentation directly from the sea-water and have been formed as a result of the rhythmic growth of the ultra-microbes.

Keywords: manganese nodule, stromatolite, Minima, Admirabilis, Actinomycetes, Miniactinomyces chinensis, Spirisosphaerospora pacifica, ultra-microfossil

INTRODUCTION

Pelagic manganese nodules have been a favorite topic of discussion among marine geologists ever since they were discovered by the Challenger Expedition in 1873. It has been known for nearly a century that the microorganisms play an essential role in geo-chemical history of manganese and result in precipitating highly oxidized minerals of manganese on the bottom of basins[26, 27]. Some scientists consider that the Mn^{4+} can be precipitated from oxygen-containing water after the Mn^{2+} was microbially oxidized[9-12, 20, 21 23 29]. However, only a few scientists have described stromatolitic features in recent manganese nodules. Farinacci (1967) and Hoffman (1969) were convinced that the columnar structure exhibited by the Recent pavement from the San Pablo seamount (Western Atlantic) are stromatolites[14, 15]. Monty(1973) considered manganese nodules to be oceanic stromatolites. He concluded that the lamellar and columnar structures exhibited by the nodules from Blake Plateau (depth 500-1100m) have been formed by filamentous bacteria concentrating Fe- and Mn-oxides[18]. However, Wendt(1974) doubted his conclusion[28]. Augustithis(1982) pointed out that the possibility that manganese nodules represent a type of stromatolite should not be disregarded and overlooked[1]. However, what microorganism is the real constructor of the manganese stromatolite remains unsolved. Therefore, the above-mentioned hypothesis were not widely recognized by most nodule researchers[2, 6, 8, 19, 25].

Two ultra-microfossils, *Miniactinomyces chinensis* Lin et al., 1996 and *Spirisosphaero-spora pacifica* Zhang et al., 1995, have been found recently in manganese nodules collected from the East Pacific Ocean at a depth of about 5000m[16, 31]. The discovery supports Monty's conclusion. Moreover, two forms of stromatolites, *Minima* and *Admirabilis*, have been identified in these nodules[4, 5]. This paper discusses the classification of the ultra-microbes, the possibility of their existence, and the implication of occurrence, radial growth of the pelagic manganese nodules and presence of cores in them.

MANGANESE STROMATOLITES

Based on the surface textures all nodules studied can be divided into smooth one and knobbly one. Both of them are composed of core and coatings. The core is usually a fragment of either preexisting nodules or rocks. The coatings are manganese stromatolite as Monty concluded in 1973[18].

The laminae and columns peculiar to stromatolites are well developed in both smooth and knobbly manganese nodules from the East Pacific Ocean. Both laminae and columns in nodules are distinguished from those in malachite, agate and some inorganic minerals. They coexist with each other and the laminae are developed within columns. The columns are developed radially from the center of nodule to the surface and can be branched and coalesced. Based on lamina shape and the style of column branching two forms of stromatolites, namely *Minima* and *Admirabilis*, can be identified in smooth and knobbly manganese nodules respectively[4, 5].

Minima
The columns in *Minima* are well developed and closely packed. They are roughly cylindrical and carrot-shaped. The diameter of these columns is about 0.1-0.25mm and the length can be over 2mm. The columns are laterally linked by bridges. Cornices but no wall can be seen on the column margins. The laminae are gently convex arch-shaped in the section parallel to columns(Fig. 1) and concentric in the section perpendicular to columns (Fig. 2). The convex laminae are always towards the surface of nodule. The style of column branching changes regularly from *Jurusania* via *Kussiella* and *Inzeria* to *Jurusania* or backward rhythmically from center to periphery

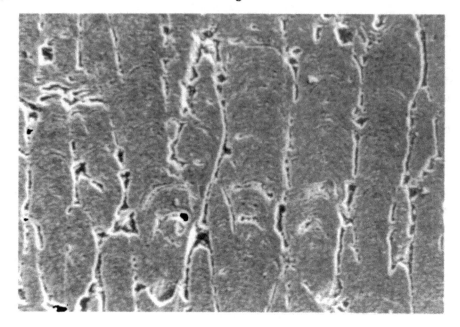

Figure 1. Polished section of *Minima* parallel to columns showing the laminae and columns.×100, 10mm=100µm.

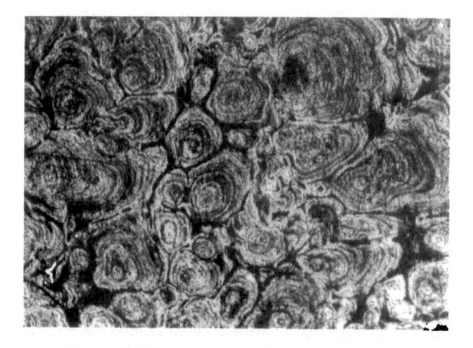

Figure 2. Polished section of *Minima* perpendicular to columns showing the laminae and columns. ×100, 10mm=100µm.

of the *Minima*[4, 5]. All these morphological features are identical to those observed in some Late Precambrian stromatolites except their size.

Admirabilis

The columns in *Admirabilis* are nodular cylindrical or tuberous. The length of columns is up to 1.5mm. They are branched as *Anabaria* with thick bud-shaped projections. The parent columns are 0.5mm wide and the daughter columns are only 0.15mm. The laminae are wavy convex arch-shaped in the section parallel to columns(Fig. 3) and concentric in the section perpendicular to columns. They are usually thick on the top of "arch" and thin on its margin. The convex laminae are approximately towards the surface of nodule. These characteristics conform with the diagnosis of *Admirabilis*[3].

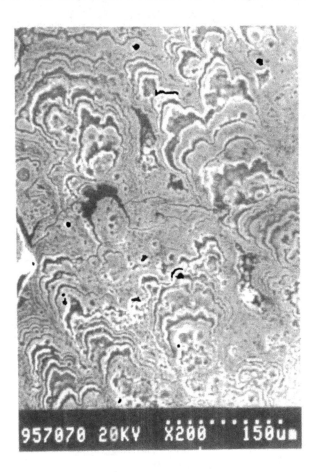

Figure 3. Polished section of *Admirabilis* parallel to columns showing the laminae and columns. ×200, 10mm=50μm.

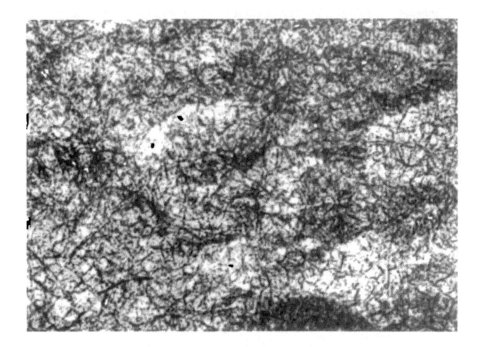

Figure 4. Electron photomicrograph of *Miniactinomyces chinensis*. ×200 000, 1mm=5nm.

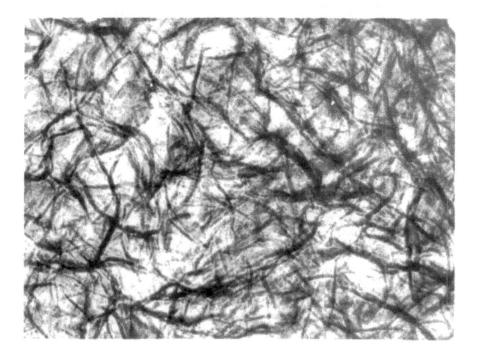

Figure 5. Electron photomicrograph of *Spirisosphaerospora pacifica*, ×200 000, 1mm=5nm.

ULTRA-MICROFOSSILS

New evidence for their microbial formation is the discovery of the spiral and chain-like ultra-microfossils by the authors in the manganese nodules studied. The spiral ultra-microfossils are named *Spirisosphaerospora pacifica* and the chain-like one are named *Miniactinomyces chinensis* Both of them are extremely small. They are several nanometers in size and about two orders smaller than common microbes. They are characterized by differentiated nutrient and generative hyphae, presence of spores, sporangia and radial growth. Except for their size, they are mostly similar to some actinomycetes in the above-mentioned features and seem to be minimized actinomycetes.

Miniactinomyces chinensis
The hyphae in *Miniactinomyces chinensis* sp. nov. are branched and interwoven with each other(Fig. 4). The spherical spores are about 3nm. They developed on the generative hyphae and resemble pearls which link each other to form a "chain". The sporangia can also be found within the network of intertwined generative hyphae. The diameter of sporangia is about 70nm and the biggest one is 120nm.

Spirisosphaerospora pacifica
The nutrient hyphae in *Spirisosphaerospora pacifica* are branched. The generative hyphae are spiral. They are about 2-4nm in diameter. The screws formed by the spiral generative hyphae are 3-25nm wide and 36-200nm long(Fig. 5). Spheroidal sporangia 9-50nm in diameter are developed in the mycelium of generative hyphae. The spheroidal spores of 4-6nm in diameter are separately preserved in sporangia.

Both *Miniactinomyces chinensis* and *Spirisosphaerospora pacifica* can be found everywhere in the coatings of smooth and knobbly nodules respectively. Consequently, they are real constructors of the pelagic manganese nodules. Some microbes found in manganese nodules, such as diatoms, radiolarias, silicoflagellates etc., are foreigners as from upper sea water or transported from else where.

DISCUSSION

Cyanophytes or Bacteria?
It is well known that most stromatolites are a result of the growth and metabolic activity of cyanophytes. What kind of microorganisms, cyanophytes or bacteria, were the constructors of the manganese stromatolites? What are the affinities of the ultra-microfossils?

The cyanophyte *Rivularia* seems to be similar to the newly found ultra-microfossils. However, most cyanophytes are photosynthetic procaryotes, the nutrient and generative trichomes are not differentiated from each other. In addition, spores and sporangia are absent in *Rivularia*, but the heterocyst is sometimes present. Consequently, these ultra-microfossils are obviously not cyanophytes.

The dimensions and some other features of different microorganisms are listed in Table 1 for comparison. The fungi are eucaryotes and their colony structure is different from that of actinomycetes. They are much larger than actinomycetes[7, 13, 22]. Therefore, these ultra-microfossils are obviously not a fungi either.

Some bacteria such as *Streptococcus* are arranged as a chain, but they are not branched and filamentous. The bacteria disperse in colonies without special colony structure. The viruses, bacterial viruses(bacteriophage) and viroids are not independent organisms. They are unable to reproduce outside living cells. Some tiny procaryotes such as chlamydias and rickettsias are obligate parasites. The mycoplasmas are probably the smallest microorganisms capable of autonomous growth on the nutrient medium. They do not form branched filaments and are highly pleomorphic because of their lack of rigidity. Therefore, the ultra-microfossils are neither bacteria nor parasitic viruses or procaryotes. They are most similar to some actinomycetes except in size and seem to be minimized actinomycetes. Consequently, the authors tentatively place these ultra-microfossils in the actinomycetes. They are a kind of sessile benthic microorganisms living in complete darkness on rocky or other solid substrates.

Existing condition of the ultra-microbes
Submersible observation, sampling, submarine photographing and video-recording show that the deep sea floor is not a desolate place. The specific Mn^{2+} oxidizing bacteria in laboratory culture grow in high hydrostatic pressures (340 to 470 bars) at 4°C which prevails in deep sea floor where the nodules occur. It has been reported that a deep-sea barophilic bacterium has been isolated from a deep-sea sample and has been found to grow optimally at about 500 bars and 2 to 4°C[30].

It has been reported that there are rich biological resources on the deep sea floor and the amount of species runs up to 10-100 million which is several hundred times the admitted number of 200 thousand. All these newly found organisms roost on or in the ooze at a depth of 1900m.

Some unidentified worms, sea stars and fishes living on and swimming above the manganese nodules can be seen on the photographs taken by the scientists of the Second Institute of Oceanography of State Oceanic Administration. Besides, there are many remains and skeletons of various planktons in ooze. They can provide benthic organisms with sufficient nutriment. Therefore, the existence of the benthic ultra-microbes on the deep sea floor is entirely possible.

Possibility of the abiogenic formation of deep-sea stromatolites
Can the laminae and columns in pelagic manganese nodules be formed by abiologic physico-chemical process? Is biogenic formation consistent with the occurrence and the morphological features of nodules?

Firstly, the manganese nodules always occur separately and scattered at the sediment-water interface. Sometimes they may be closely packed. They are entirely distinguished from the surrounding sediments in composition, structure and texture. Both manganese nodules and the surrounding sediments are formed in the same place and at the same

Table 1. Comparison of dimensions and other characteristics of microbes.

Microbes	Dimensions	Characteristics
viroids	about 2nm	small pieces of RNA or DNA uncomplexed with any protein coat, stable self-replicating infectious entity
virus	20-300nm	infectious agent that has a genome containing either RNA or DNA
bacteriophages	head: 100nm; tail: 300nm	bacterial viruses that attack bacteria
chlamydia	300-800nm	obligate intracellular pathogens of vertebrates with an intracellular pattern of reproduction
rickettsia	$1\text{-}2 \times 0.3\text{-}0.7\mu m$	coccoid or rod-shaped cells that have a strictly intracellular existence in vertebrates.
Mycoplasma	$200nm\text{-}10\mu m$	cell-wall-less coccoid to filamentous bacteria with a pliable cell boundary.
spirochaeta	$5\text{-}500 \times 0.2\text{-}0.75\mu m$	actively motile, flexible, spiral bacteria
actinomyces	about 1μ in diameter	rod-shaped to filamentous, generally nonmotile mostly aerobic bacteria forming branching filaments and network of filaments
fungi	$20\text{-}30\mu$	monocell or multicell eucaryotic microbes, sexual or asexual reproduction, with or without septum, complete nucleus

time. They coexist with each other. The manganese nodules do not cover the floor homogeneously as the surrounding sediments. Therefore, nodule-formation is a local phenomenon. In addition, the concentration of Mn in sea-water is so low that manganese oxide cannot be precipitated or crystallized directly from sea-water. In order to produce laminae and columns individually, the concentration gradients must exist separately for each lamina or at least for each column. The ultra-microbes can satisfy this condition. Each mat may result in a concentration gradient.

If the laminae and columns in manganese nodules were formed abiologically, two immiscible solutions or fluids must be present. One is nodule-forming and the other is not nodule-forming. Moreover, these two solutions or fluids must be moving in radial direction towards each other in order to produce the convex laminae and the cylindrical columns. But how can one imagine the presence of two immiscible and interlocking solutions or fluids on the deep-sea floor?

Secondly, the presence of cores in nodules and the radial growth of the columns cannot be explained on the basis of their abiogenic precipitation or crystallization. The cores in manganese nodules vary in composition and size. They may be fragments of preexisting nodules or various rocks, or they may be a shark's tooth, bones or even a man-made object. The boundary between the core and the coatings in most nodules is discrete. All these obviously suggest that the core in manganese nodules is not a nucleus of crystallization. From the biogenic point of view, the core is necessary for sessile stromatolites and acts only as a growth substrate.

CONCLUSIONS

The pelagic manganese nodules from the East Pacific Ocean are stromatolites except for their cores in some cases. Based on the lamina shape and the style of column branching two forms of stromatolites, namely *Minima* and *Admirabilis*, can be identified in smooth and knobbly manganese nodules respectively.

These manganese stromatolites are formed as a result of rhythmic growth of ultra-microbes, namely *Miniactinomyces chinensis* and *Spirisosphaerospora pacifica*. They can be found in all parts of *Minima* and *Admirabilis* respectively. Consequently, they are the real constructors of the manganese stromatolites.

Miniactinomyces chinensis and *Spirisosphaerospora pacifica* are most similar to some actinomycetes except in their size and seem to be minimized actinomycetes. Consequently, they are tentatively allocated to the actinomycetes. It is most likely that they were sessile benthic microorganisms living in complete darkness on rocky or other solid substrates.

Submersible observations, sampling, submarine photographing and video-recording show that the existence of the benthic ultra-microbes, including *Spirisosphaerospora pacifica* and *Miniactinomyces chinensis*, at the deep ocean floor is entirely possible.

The pelagic manganese nodules are not a product of either crystallization or sedimentation directly from sea-water. Their scattered occurrence, presence of various cores, radial growth and some other features can be plausibly explained by biogenic formation.

Acknowledgments

We thank Prof. Wang Dezi, Lu Huafu Lu of Nanjing University, Prof. Fan Qingsheng and Chen Yongxuan of Nanjing Agriculture University and Shi Yaqin of the Nanjing Institute of Microorganisms, Chinese Academy of Science for their helpful suggestions. Laboratory work was supported by the Center for Materials Analysis and the Laboratory of Mineral Deposits of Nanjing University, and partly by the Nanjing Institute of Geology and Palaeontology, Chinese Academy of Science.

REFERENCES

1. S.S. Augustithis. *Atlas of the sphaeroidal textures and structures and their genetic significance*, Theophrastus Publications S.(1982).
2. V.K. Banakar, R.R. Nair, M. Tarkian and B. Haake. Neogene oceanographic variations recorded in manganese nodules from the Somali Basin, *Marine Geology* 110:3-4, 393-402(1993).
3. Bian Lizeng and Z.C Huang. On classification and paleoecological significance of oncolite and feature of non-skeletal oncolite in Ordovician, Anhui, China, *Acta Palaeontologica Sinica* 27, 544-552(1988)(in Chinese with English abstract).
4. Bian Lizeng, Lin Chengyi, Zhang Fusheng, L.F. Zhou, M.Q. Zhang, Shen Huati, Chen Jianlin and Han Xiqiu. Method and terminology for describing pelagic manganese nodules, *Jour. of Nanjing University(Natural Sciences)* 32:2, 287-294(1996) (in Chinese with English abstract).
5. Bian Lizeng, Lin Chengyi, Zhang Fusheng, Du Dean, Chen Jianlin and Shen Huati. Pelagic manganese nodules;ᵃa new type of oncolite, *Acta Geologica Sinica* 70:3, 232-236(1996)(in Chinese with English abstract).
6. Y.A. Bogdanov, O.G. Sorokhtin, L.P. Zonenshain, V.M. Kuptsov, N.A. Lisitsyna and A.M. Podrazhaansky. Ferro-manganese crusts and nodules on seamounts of the Pacific Ocean, Nauka, Moscow(1990).
7. T.D. Brock, D.W. Smith and M.T. Madigan. *Biology of Microorganisms*, 4th edition, Prentice Hall, Inc.(1984).
8. G.J.de Large, B. van Os and R. Poorter. Geochemical composition and inferred accretion rates of sediments and manganese nodules from a submarine hill in the Madeira Abyssal Plain, Eastern North Atlantic, *Marine Geology* 109:1-2, 171-194(1992).
9. G.A. Dubinina. The role of microorganisms in the formation of the recent iron-manganese lacustrine ores. In: *Geology and Geochemistry of Manganese, 3, Manganese on the Bottom of Recent Basins*. I.M.Varentsov and Gy. Grasselly(Eds.) 305-326, E. Schweizerbart'sche Verlagsbuchhandlung, Stuttgart(1980).
10. H.L. Ehrlich. Bacteriology of manganese nodules. I. Bacterial action on manganese in nodule enrichments, *Appl. Environ.Microbiol.* 11, 15-19 (1963).
11. H.L. Ehrlich. The formation of ores in the sedimentary environment of the deep sea with microbial participation: the case for ferromanganese concretions, *Soil Sci.* 119, 36-41 (1975).
12. H.L. Ehrlich. *Geomicrobiology*, Marcel Dekker, Inc. (1981).
13. H.A. Lechevalier and M.P. Lechevalier. Introduction to the order actinomycetales. In: *The prokaryotes - a handbook on habitats, isolation and identification of bacteria*. M. P. Starr, et al.(Eds). pp. 1915-1922 (1981).
14. A. Farinacci. La serie giurassico-neocomiana di Monte Lacerone (Sabina), *Geologica Rom.* 6, 421-480(1967).
15. H.J. Hoffmann. Attributes of stromatolites, *Geol. Surv. Canada*, paper 69-39 (1969).

16. Lin Chengyi, L.Z. Zeng, Zhang Fusheng, L.F. Zhou, Chen Jianlin and Shen Huati. Classification of the microbes and study of the beaded ultra-microfossils in pelagic manganese nodules, *Chinese Science Bulletin* 41:16, 1364-1368(1996).

17. K.C. Marshall. Biogeochemistry of manganese minerals. In: *Biogeochemistry Cycling of Mineral-forming Elements,* P.A.Trudinger and D.J.Swaine(Eds.) pp. 253-292. Elsevier, Amsterdam(1979).

18. M.C. Monty. Les nodules de manganese sont des stromatolites oceaniques, *C.R.Acad.Sc. Paris, Serie D,* 276, 3285-3288(1973).

19. I.O. Murdmaa and N.S. Skornyakova(Eds.). *Ferromanganese nodules of the Central Pacific Ocean,* Nauka, Moscow(1986)(in Russian).

20. K.H. Nealson. The microbial manganese cycle. In: *Microbial Geochemistry,* W. E. Krumbein (Ed.) pp. 191-221 (1983).

21. S. Roy. *Manganese deposits,* Academic Press (1981).

22. H.G. Schlegel. General microbiology, Cambridge University Press (1986)

23. R. Schweisfurth, A.G. Jung and H. Gundlach. manganese-oxidizing microorganisms and their importance for the genesis of manganese ore deposits. In: *Geology and Geochemistry of Manganese, vol. 3, Manganese on the Bottom of Recent Basins.* I.M.Varentsov and Gy. Grasselly(Eds.) pp. 279-283(1980).

24. J.X. Shi and Z.Y. Chen. Abundance of microorganisms and manganese bacteria in the sediments and manganese nodules from the Northwest Pacific Ocean, *Acta Oceanologica Sinica* 11, 385-391(1989)(in Chinese).

25. R.K. Sorem and R.H. Fewkes. Internal characteristics. In: *Marine Manganese Deposits.* G.P.Glasby(Ed.) 147- 183, Elsevier, Amsterdam(1977).

26. G.A. Thiel. Manganese precipitated by microorganisms, *Econ.Geol.* 20:4, 301-310(1925).

27. V.I. Vernadsky. *Studies on geochemistry*(in Russian), Gorgeonefteizdat, Moscow(1934).

28. J. Wendt. Encrusting organisms in deep-sea manganese nodules, *Spec. Publs int. Ass. Sediment.* 1, 437-447(1974).

29. B.R. Yan, S. Zhang and D.C. Hu. Genetic relation between microbial activity and the formation of polymetallic concretions in the Central Pacific Ocean, *Acta Geologica Sinica* 66:2, 122-134(1992)(in Chinese with English abstract).

30. A.A. Yayanos, A.S. Dietz and R.Van Boxtel. Isolation of a deep-sea barophilic bacterium and some of Its growth characteristics. *Science* 205:4408, 808-810 (1979).

31. Zhang Fusheng, Bian Lizeng, Lin Chengyi, L.F. Zhou, Du Dean, Chen Jianlin, Shen Huati and Han. Xiqiu. Discovery of the spiral ultra-microfossils in pelagic manganese nodules and their significance, *Geological Jour. of Universities* 1:1, 109-116(1995)(in Chinese with English abstract).

Milton Keynes UK
Ingram Content Group UK Ltd.
UKHW040056071024
449327UK00019B/600